AF682732

RAISING GOATS
THE BACKYARD DAIRY ALTERNATIVE

BY DAVID B. WEEMS

TAB BOOKS Inc.
BLUE RIDGE SUMMIT, PA. 17214

FIRST EDITION

SECOND PRINTING

Printed in the United States of America

Reproduction or publication of the content in any manner, without express permission of the publisher, is prohibited. No liability is assumed with respect to the use of the information herein.

Copyright © 1983 by TAB BOOKS Inc.

Library of Congress Cataloging in Publication Data

Weems, David B.
Raising goats.

Includes index.
1. Goats. I. Title.
SF383.W4 1983 636.3'9 82-19380
ISBN 0-8306-1534-2 (pbk.)

Contents

Introduction

Several years ago while young and poor, I asked an old man in my town for advice on how to make ends meet. He said, "There are two ways to go about that. Eat lots of beans or get some rabbits and goats."

I got the rabbits right away. I thought one pair would soon overpopulate the hutches I built. Several months later I was still waiting for the Malthusian rabbit population explosion to be triggered. I sold the seemingly odd couple and decided that the old man didn't know what he was talking about. Later someone told me that other people in our area had the same problem that summer. What kept the rabbits from multiplying? It had been too hot that summer for rabbits.

I got some goats. My wife, Charys, and I found that a goat can live on the hay and other feed that a cow wastes. We could now afford all the milk we wanted for those extra uses like ice cream or cheese. Goat meat became our substitute for venison—at a fraction of hunting costs.

As we worked with goats, we found that most problems in raising them can be avoided by following a few simple rules. Many goat keepers needlessly complicate life for themselves and their goats. With goats an ounce of prevention is often worth a ton of cure, and this book suggests ways to prevent both physiological and psychological problems. Don't think that goat psychology isn't important.

Having goats in the family is a little like marriage. Tempers, yours and the goat's, sometimes almost get out of hand, but life is never dull. It has never been too hot yet for our goats to breed—in their proper season, of course.

Chapter 1

Why Goats Make Sense

Goats, one of the first animals domesticated by man, are the most popular inhabitants of the world's barnyards. More people drink goat's milk than cow's milk. In recent years the number of family goat herds in the United States has grown rapidly. In one 12-year period the number of registered dairy goats increased tenfold. Why?

CHANGING TIMES

During the last decade more people have decided that the time is right to adopt a more self-sufficient lifestyle. Food was bought at the supermarket during the 1960s and early 1970's with little regard for how it was produced. The food chain consisted of growers, processors, transporters, and retailers.

Some consumers began to question prepackaged dinners that were bought and tossed into the oven. While food quality was uniform and dependable, it was often mediocre. These critics looked for an alternative source of food, one for which they could have better quality control.

Others began to distrust our food supply for more ominous reasons. They either didn't like food that had been exposed to insecticides or chemical additives, or they began to wonder about the supply system's stability. Many Americans began to question the complexity of the grower-to-consumer marketing system.

There are other good reasons for the average family to become more self-sufficient. It is increasingly obvious that government economists, who used to talk about "fine tuning" the economy, have lost whatever control they may have had. With each passing year there is more talk about the insolvency of the social security system, and the gloom-and-doom prophets argue about whether the nation is headed for runaway inflation or a crushing depression. The *Cornucopia Project,* a group concerned with the future food supply, estimates that by the year 2000 a quart of milk will cost $5.29. You don't have to subscribe to one of the end-of-the-world scenarios to see that it might be a good time to choose a more prudent life-style. If you live on a tiny city lot, this may mean growing a few vegetables to ease the strain on the food budget. If you have slightly more space, you can produce vegetables and dairy and meat products. You can choose between the cow or the goat for dairy products.

GOATS VERSUS COWS

There is a rule of thumb that you can feed five goats with the amount of feed required by one cow. A second part of this statement, perhaps somewhat exaggerated, is that the five goats can produce twice the volume of milk you would get from the cow. Even if true, you can't expect five goats to produce as much milk as two good cows. Five carefully selected goats can, with proper management, equal the production of two cows of questionable dairy quality.

Your family probably doesn't need the dairy output of a cow. Here lies one great advantage of goats for milk. You can adjust the number of dairy goats to fit your needs. You will get good cost efficiency because you aren't paying to produce surplus milk that is wasted. A small family with a cow in the backyard makes about as much sense as a single commuter driving 2-ton truck to work.

Perhaps one of the best reasons for choosing goats is they make good pets. There is no barnyard animal more playful than a young kid. Children love goats. Goats pose less of a hazard to children and adults than cows. A typical dairy goat weighs a little more than one-tenth as much as a cow and is at least 10 times as affectionate (Fig. 1-1).

It's easy to keep goat pens clean. Their manure is easy to handle, dry, and very light in weight.

Goats are much more fastidious than cows. When you handmilk cows, you must be on guard against the sudden splash of urine as the cow relieves herself during milking. After many years of milking goats, I have never seen a doe do that.

Goats have reduced requirements for space and housing. Goats are very individualistic, and goat owners often treat them more as pets than economic units to be manipulated for commercial advan-

Fig. 1-1. Goats are more affectionate than other farm animals.

tage. This habit has its disadvantages for improvement of the breeds; it's hard to reject a lovable cull.

If you take your questions about cows to the owner or manager of the average commercial dairy, you may get some answers, but they are likely to be curt. These people are usually busy, with little time for casual conversation. Goat keepers are likely to be eager to talk about goats. They may have obtained their goats for an economic reason, but nearly all the goat keepers I know are hobbyists at heart.

The special attitude that goat owners have toward their animals was easy to see the day I bought my first goats. The woman who owned the goats was moving and couldn't take them with her. When I arrived to look at the goats, she went to the backyard fence and called, "Here, babies." The goats came romping across the small pasture, bleating as they came. I bought the goats, then loaded them into the back of my station wagon. I asked their names before leaving. She told me. Then, pointing to one, she said, "I just called her Nannie. But that doesn't mean I didn't love her, too." Then she put her hand to her face to hide the tears. I have since found that people often feel that way about their goats. People are not as emotionally attached to cows.

GOAT TERMINOLOGY

The woman who called one of her goats "Nannie" was violating one of the rules held by serious goat breeders. People who are trying to advance the cause of dairy goats take offense to such terms as "Billy" or Nannie." That kind of loose language, they say, downgrades goats. They prefer to use specific textbook words to classify animals (Table 1-1). The parts of a goat are shown in Fig. 1-2.

MISCONCEPTIONS ABOUT GOATS

Much of the folklore about goats is based on careless observations and even more careless transmission. Some of the misconceptions are true under certain conditions. Others are just farfetched. Following are some of the most common ones.

"Goats Stink"

This idea was the basis for a verse that poked fun at air freshener materials produced just after World War II. Those products were based on the use of chlorophyll, which was supposed to replace household odors with those of the great outdoors. The verse went something like this:

The goat that reeks on yonder hill
Has browsed all day on chlorophyll.

The pseudosophistication of this jibe is evident to any goat owner. The only goats with an offensive odor are bucks, and they are

Table 1-1. Goat Terminology.

Term	Definition
Buck	Male goat
Colostrum	The first milk produced by the dam after kidding
Dam	Mother
Doe	Female goat
Gestation Period	Time of pregnancy, about 150 days
Grade goat	A goat with one purebred parent
Kid	Young goat
Kidding	Giving birth
Lactation period	Time of milk production after a pregnancy
Purebred goat	A goat whose sire and dam are registered
Registered goat	A goat who is registered in its own name
Scrub goat	A goat of unregistered or unknown ancestry
Sire	Father
Wether	A male goat that has been castrated

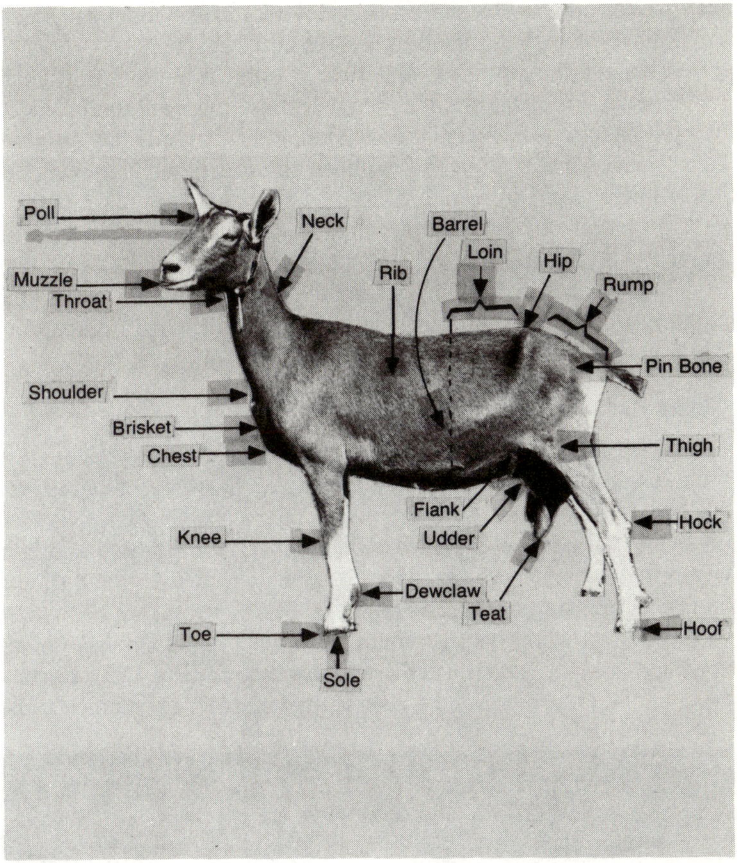

Fig. 1-2. Parts of a goat.

obnoxious only during the breeding season. This characteristic is one argument, among others, for not keeping a buck. Much of the buck's odor is produced by glands in the head, and these can be removed from male kids. This idea is pure nonsense for healthy, well-tended does.

"Goats Will Eat Anything"

This misconception probably started from observations that goats prefer different plants from those favored by cattle or other farm animals. If you put a goat in a pasture with cattle, the goat will probably begin by eating some of the weeds or other broadleaf plants. Grass will be left to cattle. Goats prefer the leaves of woody plants over others. Because of their love for brush and small trees, goats have often been used to clear land of sprouts.

This misconception leads some people to believe that goats will choose tin cans as food. Goats will sometimes eat the labels off cans. Goats are fastidious eaters. When offered moldy hay or grain, a goat will pick at it, then nod her head vigorously while moving her lower jaw up and down to dislodge the bad food from her mouth. A goat will grab a mouthful of fresh hay and pull it free with no superfluous movements.

This mistaken belief about goats can cause problems when a new owner tries to feed goats an improper diet. Goats respond to good feed and care much like any other animal.

"Goats Never Eat Poisonous Plants"

This misconception appears to have some truth in it. Goats that are allowed to roam in a wooded area are subject to poisoning less frequently than other farm animals. One probable reason is that goats will take a few bites from one plant, then move to another. Goats that are kept in a small lot don't enjoy this luxury and should be protected from poisonous plants.

One man who says that goats know what is poisonous cited as proof the fact that goats usually avoid *pokeweed* (Fig. 1-3). Certain parts of the pokeweed plant, especially the root, are known to be

Fig. 1-3. Goats are fastidious eaters.

Fig. 1-4. Does this picture prove that "goats are mean"?

extremely poisonous. Goats avoid the young shoots of this plant as much as they avoid other parts. Humans use the shoots as a wild green in dishes of poke sallet. Goats simply don't like the taste of poke. It is an acquired taste. Goats sometimes eat the ripe berries with no ill effect.

Sometimes this misconception is mixed with the one that says goats will eat anything, along with the claim that nothing hurts goats. It probably got started by someone who noticed that goats will eat poison ivy with no apparent discomfort.

Goats are somewhat less likely to be poisoned than other animals, but they can be poisoned in various degrees, including death. I will deal with specific poisonous plants and other materials later.

"Goats Are Mean"

This slanderous remark is qualified by some accusers who say, "Billy goats are mean." Figure 1-4 shows a band of goats being approached by a stranger to them at the time. Anyone who thinks that goats are mean might cite this photograph as evidence. Would you? If you really analyze the picture, it's easy to see that such a conclusion cannot be justified.

The first clue is in the man's stance. He is obviously aware of the buck's presence as shown by his outstretched hand to hold the buck away from his body.He is so unconcerned that he looks at the band of goats on the opposite side of the brook. He holds the buck at bay with nothing more than the tips of his fingers. The buck appears to be challenging the man, but he does so in a way that suggests he knows he is expected to do something. He is fulfilling the law—the one that says a male must protect his harem. This picture really shows how benign a buck can be even under circumstances that should provoke a bad actor. This is how one buck behaves with strangers, but it is the rule with many goats.

Goats with horns can be dangerous to people and to other goats. The danger is that a playful goat can hurt someone even though he had no intention of doing so. Goats that are raised in close contact with people are almost invariably affectionate. One man explained the relationship between himself and his goats by saying, "When they are babies, I'm their mama, feeding them from a bottle. When they grow up, I'm their kid, milking them."

Some goat keepers prefer to remain somewhat aloof from young bucks on the theory that making them pets causes problems. Most bucks are harmless, especially if they are dehorned when young.

MISCONCEPTIONS ABOUT GOAT MILK

An analysis of goat milk versus cow milk fails to reveal any significant differences in composition. Goat milk can be tolerated by many babies who cannot take cow milk. The reasons for the superiority of goat milk for many people with digestive troubles is not fully understood, even by nutrition specialists. Because of the almost magical cures wrought by goat milk, there are several widespread misconceptions about it. Others have been fostered just because the milk comes from an animal that is the subject of misconceptions.

"Goat Milk Tastes 'Goaty' "

This belief is based on observation that some goat milk develops a strong taste, particularly if kept for several days or longer. There are several causes for such a taste, and most of them apply equally to cow milk. How often have you heard someone say that a certain sample of milk tasted "cowy"?

Strong tasting milk is often caused by the doe's diet. Some weeds such as wild onion or garlic influence milk taste. Too many concentrates and too little bulk in the diet is another possible cause. Unclean milking practices are another cause. *Lipases,* the natural enzymes in the milk, break down the fats with time and change the flavor. This latter process can be stalled by pasteurization.

Many goat breeders believe there is a difference in milk flavor between the various breeds. Some specific comments will be included in the discussion of breeds, but there is also a difference in the milk produced by various individuals. The same differences are true of cow milk. Regardless of what causes off-flavor in some milk, whether it is produced by a goat or a cow, the best quality milk of the two animals is indistinguishable by taste.

"Goat Milk Is Richer Than Cow Milk"

It's easy to see how this misconception got started. If you have been using milk obtained from a supermarket and change to goat milk, you will be convinced that goat milk is richer. Commercial dairies limit the fat content of fluid milk to the percentage required by law for "whole" milk. The butterfat is the most valuable part of the milk, from an economic point of view, so the dairy isn't about to leave surplus fat in the milk. You are getting true whole milk with milk from your own goat, which is almost always richer than commercial milk. Goats vary in the percentage of milk fat that they produce from doe to doe, and somewhat from breed to breed, just as cows do.

"Goat Milk Is More Fat Free Than Cow Milk"

Some misconceptions about goats and goat milk are contradictory. This one is based on what many consider irrefutable evidence. If you place a bottle of goat milk and a bottle of unhomogenized cow milk side by side in your refrigerator for a day, you will find a layer of cream at the top of the cow milk but almost no cream on the goat milk. A butterfat test will show that the two samples of milk can have the same percentage of fat.

The cream in cow milk separates easily, while that in goat milk remains suspended throughout the milk. This characteristic of goat milk leads to the claim that it is naturally homogenized, which is true. It also leads to another misconception.

"Goat Milk Has Smaller Fat Particles Than Cow Milk"

This idea seems to follow from the fact that the fat particles stay in suspension. The fat rises in cow milk because the particles tend to clump together and form aggregates. The fat particles in goat milk don't clump. The cause of the clumping in cow milk is agglutinating protein absent from goat milk. By adding a small amount of skim cow milk to goat milk, you can hasten cream separation.

You would simply add some cow milk to get the cream for the butter. The milk supplies the proper protein and causes the fat particles to clump and rise.

ADVANTAGES OF GOAT MILK

Milk has long been recognized as one of the most complete foods, but some people can't tolerate cow milk. The greatest advantage of goat milk over cow milk has been appreciated by many families with babies who couldn't digest milk bought at the supermarket. The typical reaction is chronic diarrhea and symptoms of severe allergies. When these children are switched to goat milk, the cure is often dramatic. Adults who suffer from chronic indigestion or other allergies often improve with a change to goat milk. Such allergies affect people of all ages. The cure is more miraculous with babies, because they are so dependent upon milk or milk products.

I used to do some of the grocery shopping for an aged relative who has severe allergy problems. She existed on a meager pension but always insisted that I buy goat milk for her, even though it was about twice as expensive as ordinary milk. She had found by bitter experience that she must avoid cow milk.

Why is goat milk so much better for allergy-prone people? An analysis of goat milk shows that it differs from cow milk in only minor ways. This is one of those situations where science can make measurements in the test tube but can't fully explain what happens in the human body. Skeptics have offered an explanation; people who switch to goat milk expect to get better, so they do. This assumption that goat milk is nothing more than a placebo fails to explain why babies often make a rapid recovery with the change. You can hardly expect a baby to know when his formula has been changed or that he is supposed to get better by drinking goat milk.

The main measurable difference in goat milk is that the small fat particles remain dispersed in the milk, and it produces a softer curd. This apparently makes it more digestible for many people.

One advantage in producing any kind of milk at home is that you have complete control over milk quality. If you are skeptical of commercial dairy practices for any reason, get a goat and produce your own.

PEOPLE WHO SHOULD OWN GOATS

Goats are the obvious answer if you have any allergic reaction to cow milk. Commercial goat milk is not available in many neighborhoods, so it makes sense to get a goat.

It is probably easier to say who shouldn't own goats than who should. If you expect to plunge heavily into goat raising with a goal of an immediate profit, reconsider before you plunge. Most people who have made a commercial success with goats have begun on a small scale and grown gradually. Even with expert advice goat raising requires some practical experience before doing it as a full-time occupation.

One of the most successful goat keepers I know planned to stock his new farm with cattle, but he got some goats to help clear the land. As he watched the goats do the job while they existed and multiplied on a fraction of the feed needed for cattle, he switched his plans to a goat dairy. After about 10 years he has a herd of more than 100 milk goats, with an ultimate goal of 200, and a fine brick home. He is fortunate to live near a plant that processes goat milk, so he doesn't have to market the milk himself. Without such an arrangement it is more realistic to restrict your goat herd to the size that will supply your family.

If you sell raw milk, it must be to people who are buying it for animal consumption. A pasteurization plant would require a major investment. Obviously you would need to check the local regulations on milk sales, but the restriction against raw milk is probably universal.

Another source of legal problems is your local zoning laws. Typically these prohibit the keeping of animals within well developed areas of a town or city, but they can often be circumvented if you own a large enough plot of ground. The required area may vary according to local codes, but it will probably be at least 1 or 2 acres. Even if you can't meet that requirement, you can sometimes ignore these laws if no one complains. If any of your neighbors have a horse or other farm animal, there is a good chance that you can keep a goat. You would be foolhardy to invest considerable money in housing or fencing and then find that a neighbor doesn't approve of a goat in the block. The investment would make sense only if you can visualize another use for the building and fence or if it would add to your property's value.

If you are renting, remember that you would be adding value to the owner's property and that permanent additions go with the real estate. You have the owner's approval to obtain as a renter. These problems are often rather intangible ones that you can face best playing it by ear. Patience and diplomacy will often win out. Don't push your luck by attempting to keep a buck in a tight neighborhood.

Goats are natural climbers and require good fences (Fig. 1-5).

Fig. 1-5. Goats are natural climbers.

Do you have the tools and ability to build them? Check the building codes in your area and see if you can meet them without investing too much for a goat shed. In rural areas this poses no problem. If you can use a hammer and saw, you can make your own structures. Some people who keep goats hire laborers to build sheds and fences.

What is the availability of hay and other feeds in your area? Do you have transportation? A car will do, if you don't mind getting hay leaves and stems in or on it. Will the cost of driving to the nearest supply source add significantly to the price of the hay?

Do you like animals? And do you have enough patience with them to accept a certain amount of disobedience? Even if your responses to all these questions are yes, that doesn't mean you will achieve success with goats.

PEOPLE WHO SHOULD NOT OWN GOATS

At a recent fair I watched an exhibitor who had obviously invested plenty of money in some registered dairy goats. He might have spent the money on sports cars or other pursuits with more success. His goats got my attention before he did. The goats were in a pen near the show ring. As soon as he walked away from the pen, the goats left behind began to fight. He came back and slapped one goat several times. The goats ran in circles until he left, then began to fight again. When the man got to the middle of the show ring, he began to wrestle with the goat he was leading. He gave the goat a couple of hard slaps to make it behave. At the same time I noticed that the goats in nearby pens, representing other herds, were calm by comparison. There were a few bouts in other pens, apparently when a goat got out of its usual place in the pecking order, but peace was the rule. The fighting and running in circles prevailed back at the first pen. I noticed that the owner of the fighting goats appeared to be nervous in the show ring. When a goat didn't stand exactly right or moved when it shouldn't have, he cuffed it with his free hand. When the show was over, his clothing was soaked with sweat, perhaps because it was a warm day but more likely due to tension. If you see yourself in this situation, think twice about getting goats. Goats respond to gentle handling by increased production. They often react with reduced milk flow and even premature drying up when treated harshly.

If you hate to be tied down, goats may not be for you. When a doe is milking, she must be milked twice a day at regular intervals. You must be home, or arrange to have someone there, for most days of the year. Even when the doe is dry, she must receive feed and water. If you consider children in the family to be great burdens, you probably won't like goats, although the responsibilities are different.

THINGS TO EXPECT WHEN RAISING GOATS

What can you expect when raising goats? One thing is an interesting experience. Plus, you will have lots of fun.

The suggestion made earlier—that you must be willing to accept a certain degree of disobedience—may have been misleading. Rather than accept disobedience, you must be able to predict attempts at disobedience and plan how to thwart it. While goats love routine and can be trained to fit into the routine of your choice, they will welcome any opportunity that you permit to misbehave. Here is an example. Recently while rubbing the head of Buffy, a young doeling, I noticed that she jumped as if I had hit her. We were in crowded quarters, so I moved to look at her more closely, thinking she might have a sore ear. I reached for her head. Pointer, an older doe, bit Buffy on the ear opposite from me. Buffy ran away, leaving jealous Pointer to rub against me, asking to be petted.

Here's another example. When feeding several goats in a common feed trough, Myrtle takes a quick bite, then drives Pointer away. Pointer returns as Myrtle reaches for another bite. Myrtle then butts Pointer again. This routine goes on until Pointer gives up and leaves. The solution: either separate pens or a feeder that makes such behavior unprofitable for Myrtle.

Here's another example. Mandy is trying to wean her kids, which I have permitted to run with her until they are large enough to thrive without milk. Mandy pauses for a bite of grass, and each kid grabs a teat. She moves quickly to dislodge them, then sees Myrtle's kids ahead of her. She rushes them, butting the unlucky pair out of her way and leaving them wondering what they had done wrong. I am reminded of the person who comes home from a bad day at work and takes it out on the spouse or the children. Goats often behave like uninhibited humans.

Goats seem to prefer regular habits and companionship more than access to a full feed rack. Pointer has the lowest rank in the pecking order among my mature does. She is part Nubian, which probably accounts for her lack of aggressiveness. When Myrtle is near Pointer, she always asserts her dominance even if only to temporarily drive Pointer from the hay feeder. I feed grain while they are separated, so Pointer will have a chance to eat peacefully. Recently Myrtle, who had access to the pasture, strayed from the pen while I was putting hay in the feeder. Pointer could have eaten her fill of hay, but instead she went to the gate of her pen and cried. When I called Myrtle back and let her into the pen with the hay feeder, Pointer ran to the feeder with her. Myrtle stopped long enough to drive Pointer from the feeder before she began to eat hay. This kind of behavior occurs every time they are together, but when Myrtle leaves for any reason, Pointer cries in sorrow at the loss of her tormentor.

For an efficient goat operation you can establish a routine that eliminates those acts causing needless harrassment. Such routine requires planning, but you will never dictate every move your goats make. Accept and enjoy those actions you can't control. You aren't going to let a dumb animal get the best of you, are you?

Chapter 2

Kinds of Goats

While growing up in the Ozarks I saw goats wandering over the wooded hills, but people rarely milked them. They were scrub or brush goats. Most of them were descended from the Spanish goats brought to this country by the early Spanish explorers and missionaries. Their purpose was to clear land. They pruned brush and tree sprouts down to the ground.

The only thought most landowners gave to care of these goats was to adjust the number of goats to fit the area to be cleared. The trick was to use enough goats to remove all woody vegetation. Sometimes the goats received no water, so they would be more industrious about chewing off the bark from larger trees to get the sap. That killed the trees.

After clearing his land, the goat owner would advertise his herd as brush goats. Anyone who bought brush goats knew that no dairy production could be expected.

In some parts of the country Angora goats are kept for their wool production. Like brush goats, they have little interest for the backyard goat keeper. Before considering what kind of goat you will keep, it's good to know something about the important dairy goat breeds.

BREEDS OF DAIRY GOATS

There are five main breeds of dairy goats in this country: French Alpine, La Mancha, Nubian, Saanen, and Toggenburg. There is a sixth breed, the *Pygmy,* but these goats are of little interest to most goat keepers. Pygmies are smaller than other breeds, but the little acrobats require higher and more expensive fences to hold them.

The most obvious difference in the breeds are color and other aspects of general appearance. Some owners make a choice of breed by appearance alone. Your goat should be physically attractive to you, and it is one characteristic that you can be sure about before you buy the animal. If you are the practical type, you probably want to know how the breeds vary as to milk production, resistance to disease, or personality.

Buying a goat is somewhat like buying a used car. Certain makes and models of cars have reputations for a high or low incidence of a particular mechanical problem. As a hypothetical example, a 1978 Zotz might have a reputation for having few or no transmission problems. If you happen to get a 1978 Zotz that someone with little mechanical aptitude and a habit of shifting into reverse while going 45 miles per hour down the highway learned to drive with, the Zotz's excellent reputation would do you little good.

To relate this example back to goats, you might buy a breed of goat reputed to have a low incidence of mastitis. If the goat was owned by someone who kept the animal in filthy, contaminated quarters, the reputation of the breed will be of little importance.

There is considerable variation in most characteristics within each breed, so individual heredity and condition are of major importance in selecting an animal. You will learn how to judge individuals in Chapter 3.

French Alpine

The first thing you notice about this breed is the erect ears. The upright, often forward focusing ears give the animal an appealing look of alertness (Fig. 2-1). Alpines are average to above average in size, with a refined look. The head is long, and the nose is straight or slightly dished.

Alpines come in more color patterns than the other breed. Their most common pattern is brown and white or black and white patches. Some Alpines are almost solid in color, usually either brown or black. Some authorities divide the Alpine breed into three color varieties: Chamoisee, Cou Blanc, and Sundgau. The *Chamoisee* is tan or brown with black markings. The *Cou Blanc* is usually black or gray near the head, shading into silver gray and then into black at the hind quarters. The *Sundgau* is black with white mark-

Fig. 2-1. A French Alpine doe.

ings. Almost any color is acceptable for Alpines except all-white. A goat that looks as if it should be classified in another breed is always downgraded by judges and buyers. All-white is considered a legititmate characteristic of Saanens and, in some animals, of La Manchas. An Alpine with Toggenburg markings would also be discriminated against. Such variations make no practical difference for the home dairyman, unless they indicate a mistake in the breeding record.

Alpines have a reputation for high production and freedom from disease. Some members show a nervous temperament, but this trait usually occurs with inbreeding.

La Mancha

The first thing you notice with the La Mancha breed is the ears or lack of them. This characteristic attracts attention—sometimes good, sometimes not so good. Marshall Scott, a young member of a midwestern 4-H goat club, can attest to that. Marshall bought a La Mancha doeling and trained her for showing. She took first place in the Junior Division at his county's annual fair. Marshall then took the animal to another fair in which the goat division was assigned to the judge engaged to judge dairy cows. This judge placed Marshall's La Mancha at the foot of the class for having "deformed" ears. Every La Mancha owner can expect the question, "What happened to your goat's ears?"

Two kinds of ears are acceptable for La Mancha does—gopher ears or elf ears. Gopher ears are no more than 1 inch in length and often much less (Fig. 2-2). Elf ears may be from 1 to 2 inches in length (Fig. 2-3). A La Mancha buck must have gopher ears to be registered.

La Manchas are of medium size and may be of any color. The nose should be straight or slightly dished, but never Roman. Roman noses are typical only of the Nubian breed and show a history of crossbreeding when seen in other breeds (Fig. 2-4).

La Manchas have a reputation for gentleness, but this characteristic depends on how the individual is raised. The general appearance of a La Mancha is something that you may have to get used to, but if you don't object to the small ears, you will find the La Mancha a good choice for small lots.

Nubian

As in the breeds described earlier, the ears of the Nubian are an important characteristic. Nubian ears should be long and pendulous, hanging below the muzzle (Fig. 2-5). If your tie or hair has ever fallen into the water as you bent over a drinking fountain, you can appreciate the problems that Nubians have with their ears. The second dominant characteristic is the Roman nose. Nubians are average to large in size. This is the most popular breed of dairy goat

Fig. 2-2. A La Mancha yearling with gopher ears.

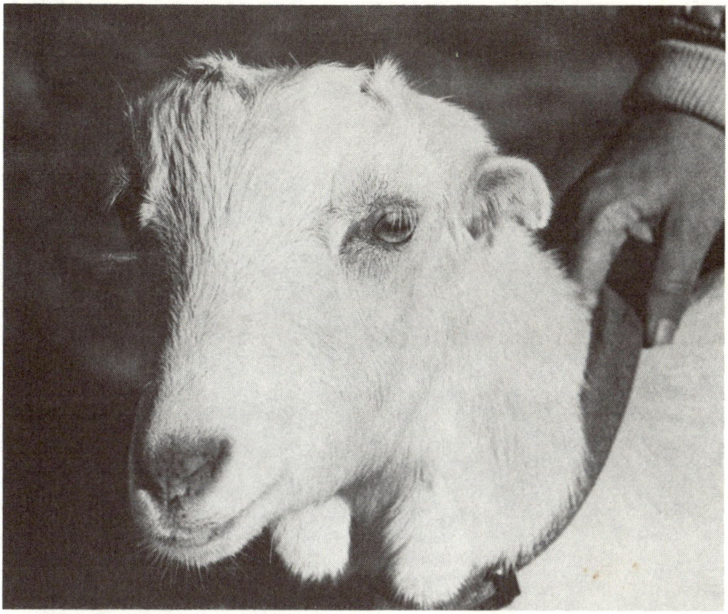

Fig. 2-3. A La Mancha with elf ears.

Fig. 2-4. A La Mancha-Nubian cross. Note the combination of elf ears and Roman nose.

Fig. 2-5. A Nubian doe.

in the United States. Nubians originated in Africa and Egypt, but were further developed in England. Nubians can tolerate heat better than the breeds that originated in Alpine countries.

In milk volume the Nubian breed is not outstanding, giving only average or below-average production. The milk is usually richer than that of any other breed. Some Nubian owners claim that the milk is better tasting than that of other breeds. It does taste richer because of the extra butterfat. Many goat dairies that specialize in other breeds keep some Nubians in the herd to raise the overall butterfat content of the milk or, in some cases, because the owner prefers Nubian milk for family use.

Some goat keepers contend that Nubians are more likely to develop pendulous udders with age. Such udders are more subject to injury than closely attached udders. Mastitis can also be a problem for udders with teats that hang low. Regardless of breed, you should look for animals with good udder attachment. If you want a Nubian for the milk quality, there is no reason to reject the breed because of the poorest members' characteristics.

One characteristic of Nubian goats that may affect your choice, if you live in a populated area, is their tendency to bleat more than other breeds. The bleat is often louder and more jarring on the nerves of some people than that of less complaining goats. Nubians seem to bleat when they see you, when you approach or leave them, and after you have left them. They bleat when it's feeding time, when it's raining, and often when the sun is shining. When several goats gather at a hay rack, there is often some jockeying for the best position. Nubians are often at the bottom of the pecking order in this situation, but they make up for it by bleating or even barking at the others. If the complaining bleat of the Nubian bothers you or if you live in a populated neighborhood, you should consider some other breed such as La Mancha.

Saanen

The most obvious characteristic of the Saanen is the all-white color (Fig. 2-6). Saanens are large animals, but they should be refined, with a straight or slightly dished nose. The ears should be rather erect, although this characteristic doesn't usually seem to be as pronounced in Saanens as in Alpines or Toggenburgs.

Along with Alpines, Saanens have a reputation for heavy production. The butterfat content of Saanen milk is only average. Some Saanen owners claim that Saanen milk tastes better than that of other breeds. It is difficult to say whether you can judge breeds on the basis of milk taste. Any rational comparison would require a taste panel for sampling the milk of hundreds or even thousands of goats. Some breeds may be more attracted to certain strong tasting

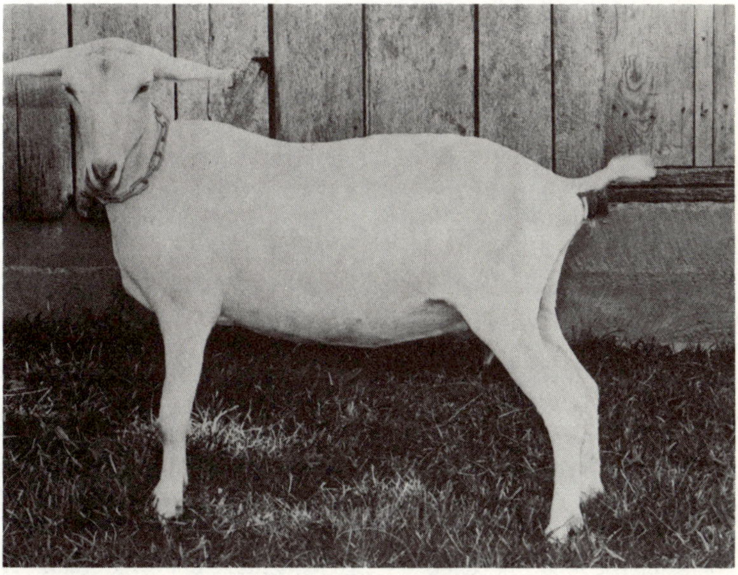

Fig. 2-6. A Saanen doe.

Fig. 2-7. A Toggenburg doe.

plants than other breeds, but this is likely to be an individual characteristic.

Toggenburg

Toggenburgs are the oldest recognized breed in the United States. They are easily recognized by the overall brown color and white face markings (Fig. 2-7). White also appears on the lower legs, from the foot to the knee or hock, and on each side of the tail in the pinbone region. Toggenburgs are often smaller than average in size.

Toggenburgs have a reputation for heavy production for their size. TheToggenburg is usually smaller than the Nubian breed, but the average Toggenburg will outproduce the average Nubian. Toggenburg milk is relatively low in butterfat content.

Because of its smaller size, the Toggenburg is often recommended for small lots. Toggenburgs are reputed to like grass better than the average goat, which prefers browse. Some goat keepers claim that Toggenburg milk has a different taste, perhaps because of this difference in food preference. Again, the difference in individual goats and the food they choose is probably more important than any variation between breeds.

CHOOSING A BREED

As you read the comments on the various breeds and look at the photographs, you may see one that appeals more than the others. If you are open to suggestions, some aspects of your own situation should be considered.

Find out what breeds are available in your area. If you obtain a registered or even a good grade doe of any one of these breeds, you will probably want to breed her to a purebred buck. If the nearest buck is located 100 miles or farther from your home, that would be impractical. Artificial insemination is available in some areas. Check your telephone directory's Yellow Pages under "Artificial Breeding." Such services are usually offered for cows, but some inseminators keep goat semen. When you call the inseminator, see which breeds of goat semen are available.

Another consideration is the distance to the nearest neighbor. As mentioned earlier, Nubians often are more vocal than other breeds and may irritate other residents in your area. If this is a problem for you, the La Mancha breed mightbe a better choice. La Manchas are typically calm and peaceful. After owning goats of every breed, I was impressed by the quietness of the first La Mancha goats I obtained. There is considerable difference between individuals.

If you are not going to keep purebred stock, choosing a breed is of less importance. You can even start with one breed and, by continual crossbreeding, gradually shift to another breed. The choice of an individual is much more important than the choice of a breed for most goat keepers.

Chapter 3

Getting Your First Goat

Before rushing out to buy a goat, there are several decisions to consider. Will you look for a purebred or a grade animal? Should it be a young kid or a milking doe? Where will you look? How will you know that you aren't getting a "lemon"? Failure to consider each of these questions can lead you into an impulse buy that you will later regret.

PUREBRED VERSUS HYBRID OR GRADE DOES

The choice here depends partly on your reasons for getting a goat. If you want to sell the kids at good prices, a registered purebred doe may pay for herself. She must be bred to a registered buck, which means a higher breeding fee. If no registration papers are required, you can usually breed your doe to the same buck at a reduced fee.

A good grade doe makes better economic sense for many goat keepers. A grade doe will be priced below that of a registered doe, and your breeding costs will be lower. The kids, especially the male ones, will be good for meat only, but you can use the meat to reduce family food costs. Many good grade does will produce just as much milk as a purebred. One goat dairyman I know claims that he gets better production from crossbreeds and grades than from purebreds. This is especially true when a certain line of purebreds have been subjected to intensive inbreeding. I know of one registered buck, whose dam had been a state champion, that grew up to be

Fig. 3-1. A registered buck that is undesirable because of weak forelegs.

weak in the forelegs. The buck kneeled to graze (Fig. 3-1). I learned later that the buck was the product of inbreeding, a practice that is almost necessary in certain breeds in some parts of the country. This occurs when there are not enough registered animals around to bring more variety into the gene pool. Some breeds have been developed from a very few original animals imported to this country.

If your interest in goats includes showing animals at fairs and trying to improve the breed, you must choose registered does. You can eventually own an established herd of registered animals and expect to sell the kids at a higher than average price. While this may appear to be an easy way to increase your income, there are many obstacles to overcome before you can show a profit. You must learn whether there is a market for registered kids in your area or if such a market is likely to develop. Even if the prospects look good, it is better to grow into this kind of business than to buy your way into it too quickly. Learn on a small scale, then expand.

KIDS VERSUS MILKING DOES

This kind of choice may depend on how quickly you must have milk production. If you have a sick baby in the family who can't tolerate cow milk, the choice is made for you. If you are merely looking forward to reducing your family food budget in the coming years, you can buy a kid at a reduced price and let it grow.

This doesn't always save as much money as you might think. You will lose at least several months or a year of milk production by raising a kid. You must buy feed and hay during that time. Kids are moderate eaters, but it all adds up.

Another argument for selecting a milking doe is that you can see the fully developed animal and better judge her quality. I know one dairy goat judge who refuses to place junior champions above other class champions when judging the various classes for a grand champion animal. He says that the mammary system is the most important part of the goat to inspect, and young kids just don't have enough udder development.

His comments suggest that choosing a kid may be like playing the tables at a Las Vegas casino. Kids can be judged as to dairy character, but they can also be chosen on the basis of their sire and dam. A doe with a history of producing doe kids that develop well can be expected to do so in the future. If the kid comes from a herd where production figures are recorded, you can check the records of the dam and any full sisters in the herd. Discount any records reported as gallons or quarts per day; accurate records are always kept by weighing the milk and recording the number of pounds. The most reliable records are those certified by associations such as the DHIA (Dairy Herd Improvement Association), but records for goat herds are rare.

The herd owner is likely to sell his poorest does first. You probably have a better chance at getting a good producer by buying an untested kid. This would be especially true if the owner were selling all his doelings. When you find which animals are for sale, check each one against the list of faults to avoid (Table 3-1).

SOURCES OF GOATS

Before buying a goat, make a list of every local source, the breed and type of animals available, and any special comments that apply to that herd. To find the names and addresses of local herd owners, ask your county agent or farm advisor if there is a local dairy goat association. The officers in such organizations can furnish a list of members.

Another possibility is to check with your local 4-H club leaders about any goat clubs in your county. The best 4-H source of information on goats will be the county goat leader. In my county it is Ruth Scott, the mother of Marshall Scott, the boy who showed the La Mancha in a fair with an ignorant judge. Ruth grew up in a large city, then moved to a small town where she became interested in 4-H work and, eventually, goats. She knows the location and quality of

Table 3-1. Checklist of Defects Serious Enough to Warrant Rejection of an Animal.

Causes for Rejection—Milking Doe

Udder Problems

- Mastitis
- Extra teats
- Teats too large or too small
- Blind teat or double orifice
- Pendulous udder
- Hard Udder

Poor Health

- Blindness
- Lameness, except for temporary injury
- Swollen joints
- Hernia
- Abcesses
- Weakness or emaciation

Causes for Rejection—Doeling

Symptoms of Hermaproditism

- Masculine appearance and behavior
- Unusual sex organs
- Naturally polled

Udder Defects

- Extra teats

Poor Health

- Same criteria as for mature doe

every goat herd in her county and many of those in surrounding counties. Her knowledge of goats is well known, so people often come to her for information on where to obtain goats, to have does bred to a suitable buck, or to sell them. County 4-H leaders are volunteers who work without pay, so they are much more likely to have the time and enthusiasm to answer your questions than is a commercial goat dairyman.

If your county has a fair that includes a goat division, contact the people who show goats there for information on where to buy animals. Fairs are good training schools, because the judges who rate the individuals in each class usually tell why they placed each goat as they did. Most fair judges are competent, unlike the one who downgraded a La Mancha for having "deformed ears."

Local newspapers often carry ads about goats for sale in the want ad section. Throw-out sheets are often the best source of ads. You can get these free at your local supermarket or other stores. They contain no news—just advertising.

Sale rings such as stockyards are a questionable source. You will have little opportunity to carefully judge an animal at these places, and there is the increased hazard of disease transmission.

Try to find someone who must sell an entire herd. If you get there first, you can have the choice of the herd. No one knows the positive and negative characteristics of each animal as well as the herd owner. If you must buy from his surplus, you are getting the individuals that he has culled. If you do buy from a herd surplus, your most likely source, carefully examine the animals offered for sale so that you will choose the best one available. Don't forget to use Table 3-1 to eliminate problem animals.

CHOOSING A DAIRY GOAT

As mentioned earlier, it is much easier to judge a milking doe than a young doeling whose udder is undeveloped. The udder is the single most important part of the animal to examine, but even with kids you will want to choose a healthy animal that has no obvious problems. If you are buying a kid, you can ask to see the dam and compare her udder development to the ideal. Following the udder examination, inspect the entire body for defects.

Udder Quality

The ideal udder is large enough to indicate good production but with extended forward attachment and a high, smooth rear attachment (Fig. 3-2). Many large udders are undesirable. The most obvious kind is the pendulous udder that hangs so low that the floor of the udder is at a lower level than the doe's hocks. These udders have poor attachments, often with a shape like a ditty bag with the drawstring pulled tight (Fig. 3-3). Looking at the udder from the side, you should see a smooth transition from the udder to the body (Fig. 3-4) instead of a pocket there (Fig.3-5). A poor udder will often appear to be divided into sections (Fig. 3-6). The teats should be of uniform size and shape, with a slight forward angle (Fig. 3-4). Reject full-grown does that have small udders or teats (Fig. 3-7).

Inspect the doe from the rear to check the position of her hocks. This can be an indication of quality even if the doe is dry. The legs should be widely spaced to make room for a capacious udder (Fig. 3-8) instead of close together (Fig. 3-9).

If the doe passes these tests, gently feel the udder for hard places. Carefully inspect the udder skin for sores, warts, or scars. Unless the doe is dry, you can check the teats for proper orifices by squeezing out a little milk from each teat. The milk should flow freely; if it is stringy, suspect mastitis. Watch for double orifices or

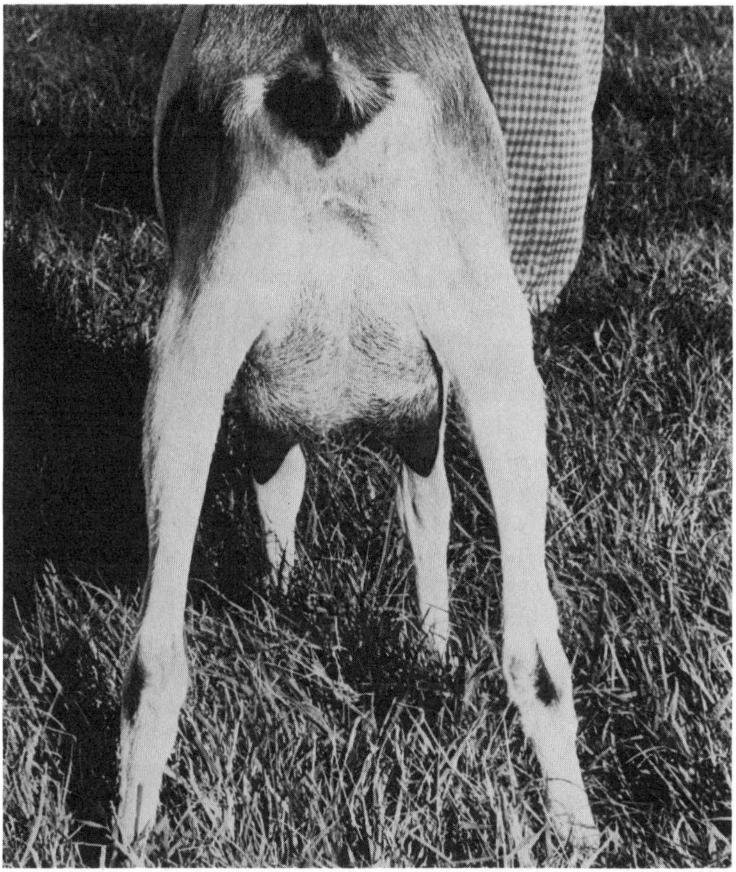

Fig. 3-2. A good udder with broad, smooth rear attachment.

extra teats—reasons to reject a doe. Some buyers like to taste the milk at this time to see if it has any off-flavor characteristics. Make sure the animal is healthy. A set of healthy kids is some indication of a healthy doe, but no guarantee. Many people drink raw milk from cows or goats with no thought of disease; others insist on pasteurization. Decide for yourself. One man I know wants his milk pasteurized, so he simply smells the milk from a prospective doe. (Milk for babies should be pasteurized, because babies are susceptible to bacterial infections.)

General Appearance

The doe should be an alert animal, with head held high. Avoid a young doeling that has an exaggerated barrel but is poorly de-

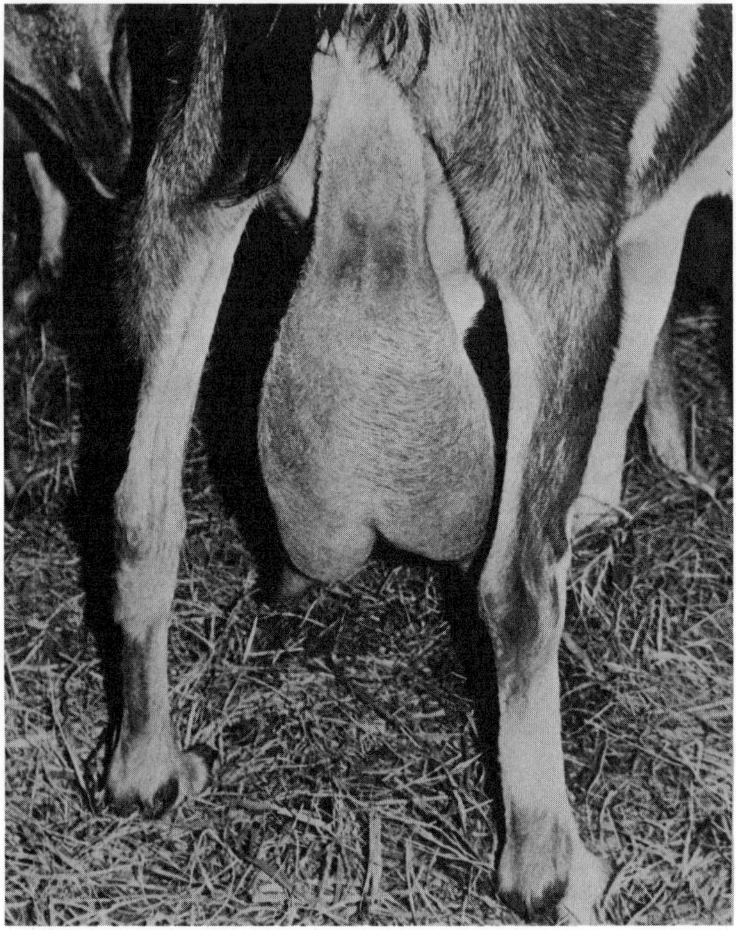

Fig. 3-3. A poor udder—too pendulous and with narrow attachment.

veloped in other ways. This condition is known as *pot belly*. A large barrel that blends with the rest of the body is desirable and indicates a large rumen. The back line should be smooth with no deep depressions and a slight slope at the rump. The front legs should be well spaced to permit good body development (Fig. 3-10). The joints should be smooth. Reject any animal that has enlarged joints (Fig. 3-11).

Head

The ears should be normal for the breed. The nose should be straight or slightly dished, except for that of the Nubian breed. Look for a broad muzzle and large nostrils. The neck should be refined, blending into the shoulders. Reject any animals with coarse or crooked features.

If the animal comes from a horned herd, it should have been dehorned while young. *Scurs,* the deformed remains of improper dehorning, reduce the value of an animal. These often cause problems by breaking off periodically, leaving a bloody stump. A goat's horns can cause injury to other goats or to you. If a horned goat throws its head up while you are bending over it, the result can be painful.

Check the teeth to estimate the age of a mature goat. Look at the temporary teeth of young kids. Then ask to see a yearling. The yearling will have a pair of permanent teeth, the first incisors, at the center front of the lower jaw. Then she will add a pair a year until she

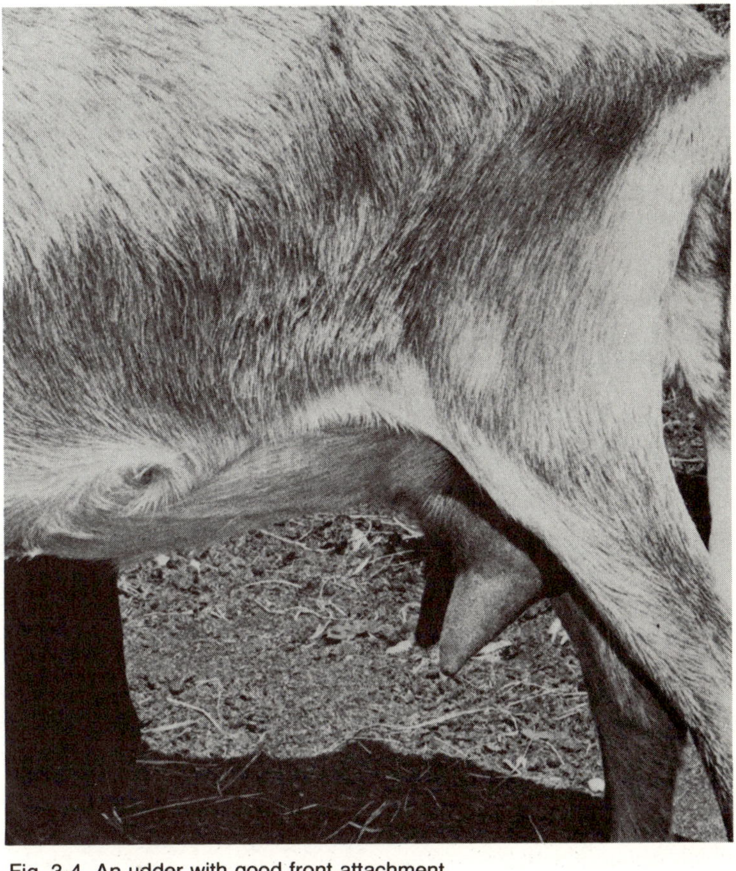

Fig. 3-4. An udder with good front attachment.

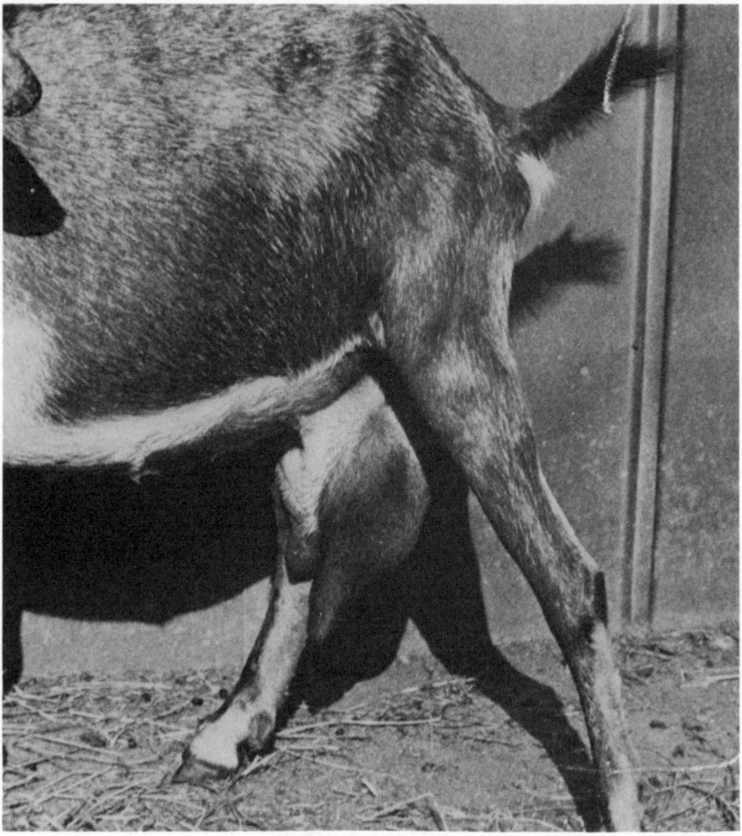

Fig. 3-5. A poor udder with a pocket at the front attachment area.

is four years old (Fig. 3-12). These incisors replace the small, pure white baby teeth. After the fourth year there is a gradual increase in the spaces between the teeth, and the teeth become more worn and irregular. A goat has no teeth at the front of the upper jaw. You can learn to estimate the approximate age of an individual by the number and condition of the incisors and other teeth.

While examining the head, look for open sores, old scabs, or a discharge on the ears. Feel under the jaw for lumps. Some goat herds have problems with abscesses that appear as lumps, often near the lower jaw. Such abscesses often show drainage, perhaps because the owner has lanced them (Fig. 3-13). Abscesses can be contagious, so you should reject any animal that has one or more. Avoid animals with hernias; a hernia can become strangulated, which causes death.

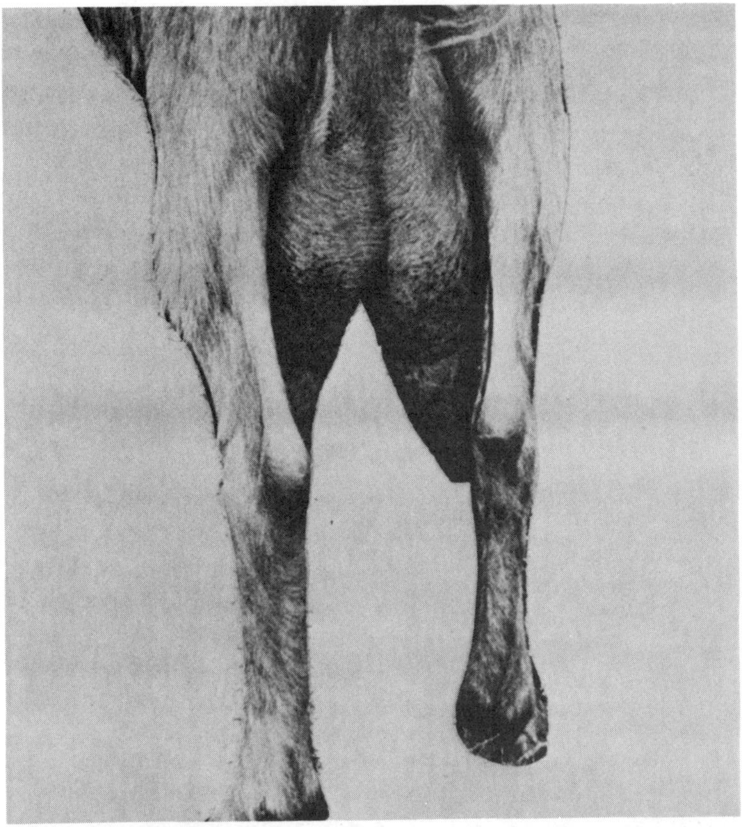

Fig. 3-6. An udder with too much cleavage.

Fig. 3-7. Reject does with small teats. They are hard to milk.

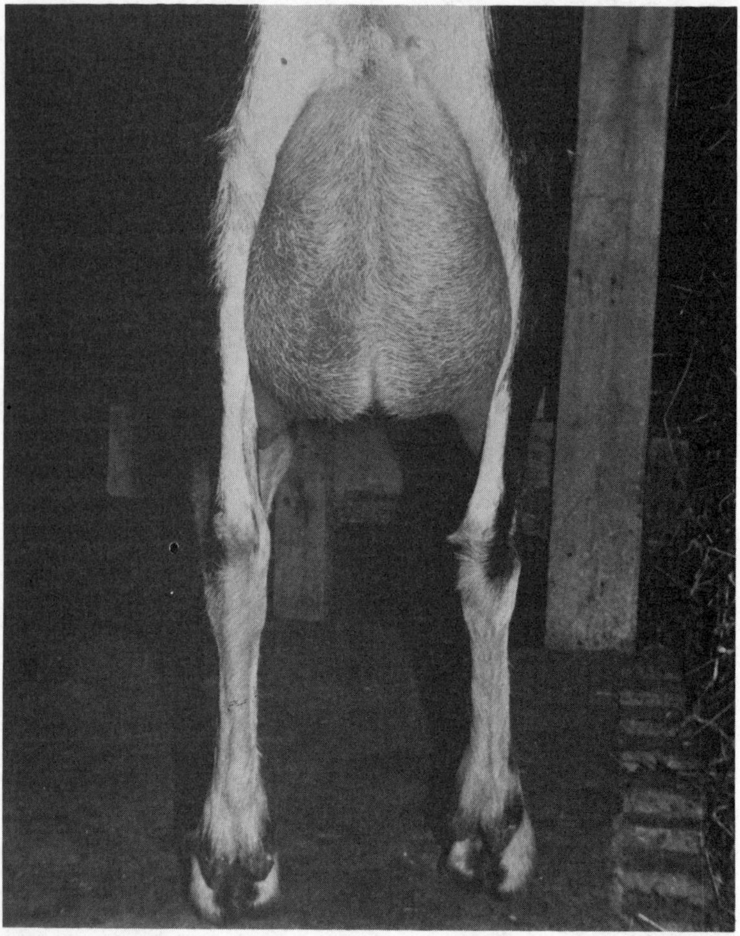

Fig. 3-8. Look for widely spaced rear legs that make room for a capacious udder like this one.

Watch the doe as she walks. A doe that limps should be rejected. A limp can indicate a temporary injury, but it may mean foot rot or another serious problem. See if the doe stands well or if the feet appear to be turned or broken down.

Make sure her skin looks healthy and the coat is glossy. A rough coat usually indicates some health problem, but it may not be serious.

Dairy Character

One of the most important indications of good dairy character in goats is *angularity.* A doe with good dairy character will appear to have greater body depth and width at the rear than at the front. The body should be long with widely spaced ribs. Run your hand over the rib cage to see if the ribs are flat (good) instead of round. The neck should be long and narrow but smoothly blended into the shoulders.

Dairy character is evident in animals that have a refined rather than a heavy appearance. A doe should be strong and well developed but not fat. There is some variation in the desirable degree of fleshiness from breed to breed. A Nubian doe will normally appear to be fatter than an acceptable doe of other breeds. To consider how fatness affects dairy character, notice the difference in a Holstein cow and an Angus. An Angus will appear to have a stocky rectangu-

Fig. 3-9. Reject does like this one with hocks that are too closely spaced.

Fig. 3-10. The front legs should be well spaced to pemit good body development.

lar body, while the dairy animal will be more streamlined. A Holstein cow with a large frame will usually produce more milk than a small one. A cow with a reasonable degree of fleshing is more desirable than one that is emaciated. If you visit a fair where quality dairy goats are judged, you will soon begin to identify the difference between a well developed dairy animal and one that is just too fat.

Feminine Quality

Avoid does that look or act too masculine. This is particularly important if you are buying an untried doeling. Certain strains of goats have the possibility of producing *hermaphrodites,* which are animals of mixed sex characteristics that will not breed. Be suspicious of any young "doe" that snorts vigorously at other does, particularly if they are in heat. Many breeders destroy hermaphrodites at birth, if they are recognized, but other goat keepers are not

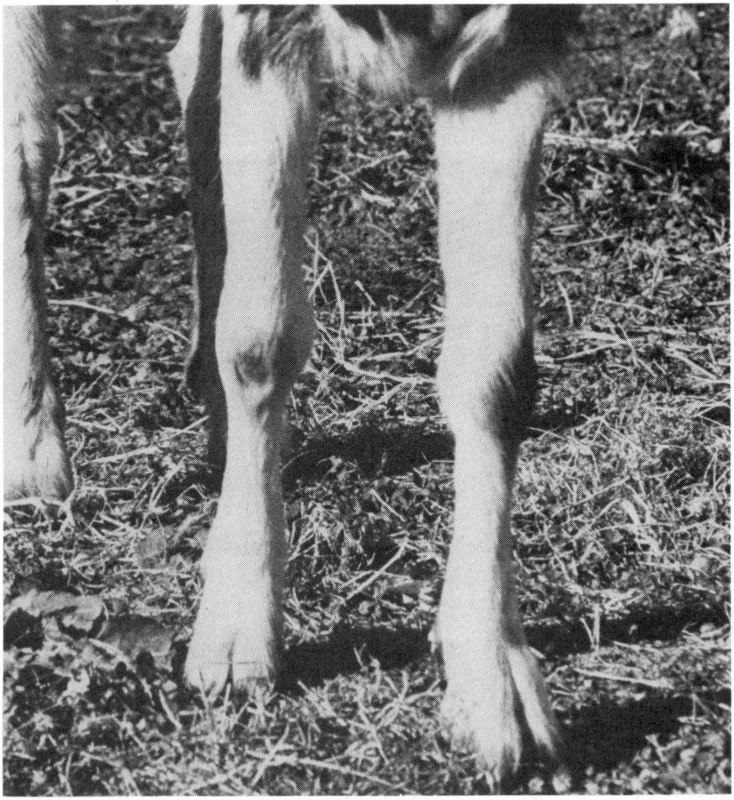

Fig. 3-11. Reject goats with swollen joints like these.

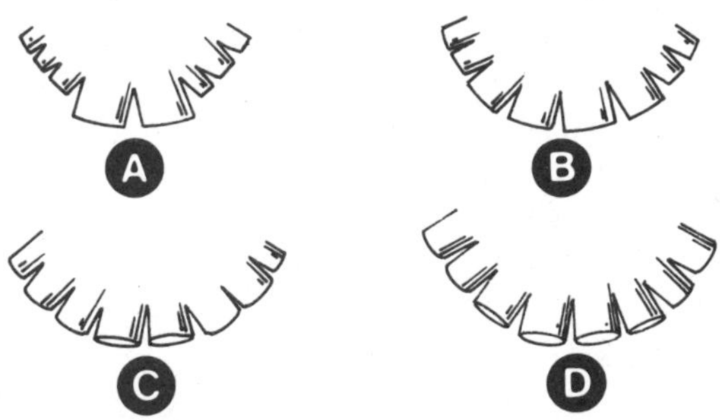

A—one year old, first pair of incisors
B—two years old, second pair
C—three years old, third pair
D—four years old, fourth pair, replacing the eight original "baby" teeth.

Fig. 3-12. Tooth development in a young doe.

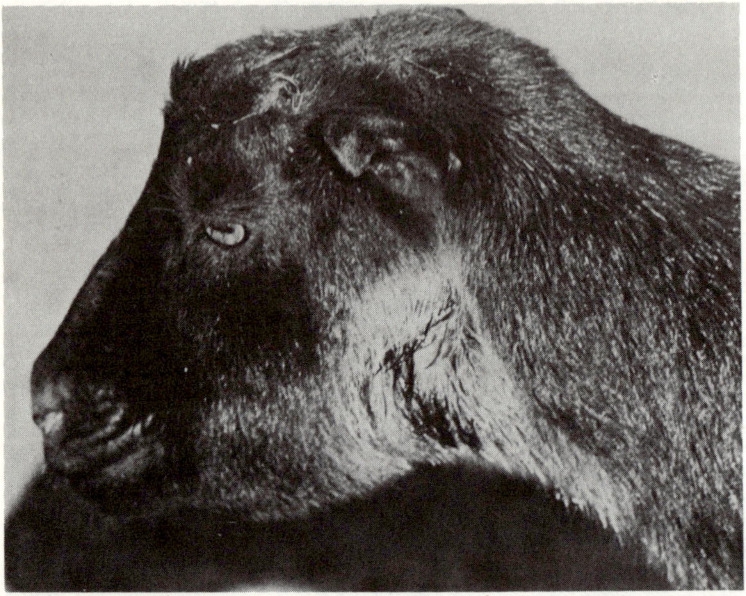

Fig. 3-13. Reject any goat that has any kind of abscess, either a lump or a running sore like this one.

even aware of the problem. They may offer one for sale in the mistaken belief that it is normal.

Hermaphroditism is most likely to occur when two naturally polled (hornless) animals are mated. Some does produce hermaphrodites regularly. When a polled buck is mated to a polled female, there is always the possibility of hermaphroditism.

Inspect the sex organs to verify the possibility of hermaphroditism. If the doeling's vagina appears to be unusual in any way, there is room for doubt. If it appears to be tipped or larger than normal, reject the animal. Other symptoms are messy urination that stains the back legs, an overly developed clitoris, or a misplaced clitoris. You can recognize the problem if you compare the sexual organs on the suspected animal with those of a normal doeling. The difference in personality of a hermaphrodite makes the problem evident in most cases.

MOVING YOUR GOAT

Make thorough preparation for easy movement of the goat and for a smooth transition from her old home to the new one. Failure to do this can create situations that will be frustrating to you and upsetting to the goat. In many cases where no planning is done, the new owner finds himself with a goat that is dropping rapidly in production. The goat may go dry in some cases. Goats are basically conservative animals. They are sensitive to change and don't tolerate radical changes or chaos.

To illustrate how some goats resist change, a friend of mine recently tossed a pad of hay to a goat in my pasture. The goat, a six-month-old wether, grabbed the hay and went running to the hay feeder in the barn lot where he ate the hay. The move may have been for security. He could have been challenged by a larger goat in the pasture, but it is also typical of how goats react to change.

Before leaving home to get the goat, check that you have comfortable quarters ready for her. A clean shelter for rainy days is needed, because goats would rather go hungry than get wet. Fences should be in excellent condition to hold her. She will be extremely unhappy for the first day or two, and that is when the fence will be tested.

Here is how one goat reacted to a strange environment. Several years ago I visited a goat dairyman who lived about 20 miles northwest of my location and bought a French Alpine doe of average size named Little Blue. I brought her home and put her in a pen alone, expecting to move some other goats in later. The next day was a holiday, and I went to a celebration at a neighboring town in the afternoon. When I returned about two hours later, Little Blue was gone. I spent the rest of the day crisscrossing the countryside for a mile or two in each direction, stopping to ask people if they had seen Little Blue. The next day I continued the search, and there was still no sign of Little Blue.

That night a friend who worked in a town five miles northwest of my location called to say that a resident of the town where she worked had found a goat. I drove to the town and found Little Blue in a pen with other goats. The finder said that Little Blue arrived on his porch Monday at 10:00 P.M. and huddled against his door to escape some dogs. She had evidently gone five miles in about seven hours or less and was headed on a beeline for her old home, although I had taken a different road to haul her home. I took her back and made sure that she had the company of other goats. I also fixed a gap in the fence that I had overlooked earlier.

This story shows one problem of keeping a single goat. Goats love company, so it's good to have another goat around even if it is nothing more than a wether that will later be butchered for meat. You can successfully keep a single goat, but such isolation sometimes makes them difficult to breed.

If you buy a doe who has had kids running with her, try to buy one of the kids. Keep it with her during the transition period.

For easy handling, take a collar and leash when you go to get the doe. In emergencies you can use a rope if you make sure it won't choke her, but a good collar is safer. You must have a leash to tie her in the vehicle for hauling.

If the vehicle is a pickup truck and the weather is cold or damp, provide the doe with protection from air movement. The chill factor is greatly increased by air movement, and goats cannot tolerate chilling.

You can use a car to move a goat. Remove the back seat and cover the floor with a thick layer of old newspapers or other disposable material. If you can force the goat to lie down during the trip, there will be no problem. I have hauled goats in sedans and station wagons without incident, but the newspapers are good insurance. If you or any member of your family objects to a goat in the back seat, you can use the trunk, but some provision for ventilation is necessary. You can remove the trunk lid completely. This converts your car to the equivalent of a small pickup and is a useful trick when you need to haul hay.

Before leaving with your new goat, ask the former owner to describe in detail how she was managed. Find out what kind of feed he used, how much, and the time of day for each feeding. Any changes that you plan to make must be made slowly. Many goats have sickened after quick changes in feed and environment, as you will notice in the case studies related in Chapter 8.

Find out the date of the last worming for the former owner's goats. Some owners do no worming at all. It is no reason to avoid a healthy animal and may even be a sign of good management practice. A recent worming or other medication means that it may be prudent to discard the milk for a while. The more you can learn about the past history of the animal you are buying, the better.

When you get the goat home, restrict her to a small pen for a day or two unless you have good fences in a larger lot or pasture. Check the list of poisonous plants given in Chapter 8, and make sure you have none that will be available to the new goat. Goats normally have few problems with poisonous plants because they eat only a few bites of each plant as they move about, but a goat that is suddenly exposed to a new taste might be tempted. If you must tether her at first, prepare a tether that will avoid problems of a tangled chain, as described in Chapter 5.

Don't be surprised if a few minor difficulties arise when you first get the goat to your place. Maybe it will be a neighbor dog bounding out to make her nervous or some other unforeseen event.

If the important preparations are right, though, she will tolerate a few nuisances.

Getting a goat is an interesting project. Sometimes the search can be almost as exciting as the final selection and purchase. Happy hunting.

Chapter 4

A Home for Your Goat

Consider your maximum herd size, whether you will choose stall or loose housing, and how much hay storage space you will need when planning a shelter for your goats. Choose a suitable site for the structure and plan for proper exposure and pen access. Goats do not require expensive buildings, but careful planning can make the most rudimentary shelter much more pleasant for both the animals and you.

LOOSE HOUSING VERSUS STALLS

Loose housing, where several goats are housed in a single room, costs much less than a barn with separate stalls for each animal. Although goats like the company of other goats provided by loose housing, they will fight for dominant position in the herd. The goats that lose sometimes eat less well because the more dominant goats drive them away from the feeders.

Another problem with loose housing is that heat buildup can cause condensation in cold weather, adversely affecting the animals' health. This is a problem only if the structure is virtually airtight. The typical homemade goat barn will have numerous air leaks that provide desirable ventilation. The air leaks on the sides of the barn should ideally be plugged so that winter winds can't blow through the space occupied by goats. If the sides are airtight, then some kind of under-the-eaves or ridge ventilation should be provided (Fig.4-1).

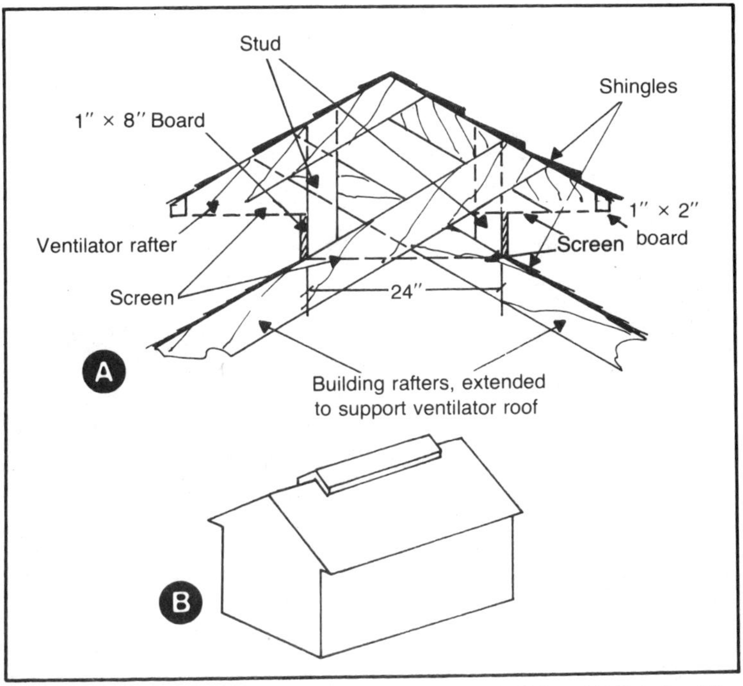

Fig. 4-1. One method of adding a ventilator to a new building. (A) Details of the ventilator. (B) Placement of the ventilator on the roof.

One of the chief disadvantages of loose housing is that you can't tailor the feeding program of each animal to her needs. You can control feeding while a doe is milking, but during her dry period you will have no control if she feeds with the rest of the dry herd. There is no problem with feeding hay; goats should have access to unlimited roughage either by pasture or hay.

Dairy goat operators frequently use houses with small stalls, some as small as 20″ × 36″. The goats are sometimes permitted to move around in the shelter—a condition known as *free stall housing.* They can be tethered to the front of an otherwise open stall for *tie stall housing.* Because the stalls are just large enough to house one goat, the space is efficiently used. People who keep only a few goats prefer to provide roomier quarters for their animals with a closed stall for each goat. Such stalls are often as large as 7′ × 7′, although half that much space is considered adequate for each goat when loose housing is employed.

Many beginning goat keepers start with open housing and alter the structure to fit their needs. Regardless of the type of housing

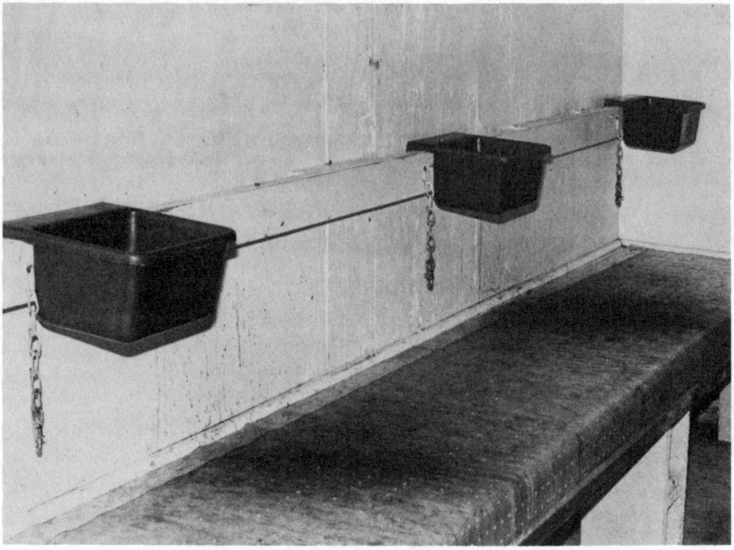

Fig. 4-2. A milking stand capable of handling three goats at once.

you choose, try to provide a separate room for your milking area. A milking stand placed where goats have access to it can't be kept clean enough for good milk quality. The size of the milking area will be determined by the number of does you plan to keep. You can get by with a single milking stand with one or two does, but for more than two goats you will probably want several stands or even a large platform with several feeders (Fig. 4-2).

FLOORS

If you are building a new structure, you may consider choosing a concrete floor for easy cleaning. Concrete has its disadvantages. It is cold, it holds urine, and it can cause udder damage if bedding is displaced. You can overcome some of the heat sink characteristics of concrete by using bedding to insulate the goat from the floor. This adds to your maintenance costs.

Dirt floors for your goats are less expensive. They are relatively warm and require no maintenance other than normal removal of manure and old bedding. A building situated in a damp location may have seepage of moisture into a dirt floor. A building in such a site will probably also have a moisture problem with a concrete floor unless suitable drainage is provided. Good drainage will protect a dirt floor.

Wood floors are rarely used for goats except when an existing structure is adapted for goat keeping. Wood can become soaked with urine. Boards may expand and warp. If the floor is in poor condition, floorboards can injure animals.

Considering advantages versus costs of each kind of floor, dirt floors are likely to be your best choice. Urine seeps down into the soil instead of pooling. Goats can be kept adequately warm in a dirt floor barn with no heating. Some goat owners install heating, but its main purpose is to make the owner feel comfortable while tending the herd. The goats don't need it.

SITING A GOAT SHED

Goats thrive best in dry locations, so build your shed on well drained ground. A slight rise is excellent. For maximum sun exposure to promote warmth and drying, face the shed to the south. Remove the topsoil from the shed floor area and replace it with rocks, topped with gravel.

If the ground around your shed location is flat, get some surplus soil from another building site. Build up the floor area until it is higher than the surrounding ground level. When cleaning out the shed, you will probably remove some floor material. This loss will eventually put the floor below ground level. A low floor will become wet and stay wet, making a shed that is messy and unhealthful.

Sometimes a building will be wet even though careful planning was done. If you find that your finished shed is wet, dig a trench around the outside walls and fill it with big rocks. Locate the direction of natural drainage and dig a trench in that direction, again using large rocks to fill the trench. Cover the rocks with surplus asphalt shingles or other durable material, then top off with soil. The open spaces between the rocks will provide easy drainage for moisture that would otherwise collect in the shed's floor.

BUILDING MATERIALS

Successful goat barns have been built from materials such as concrete and steel, but inappropriate items are sometimes included in the construction. Plywood is a useful material for homes, but it is usually bad for goat barns. When kept inside during bad weather, goats get bored and start pulling at loose pieces of wood. Sooner or later they will begin to strip the layers from a plywood wall, exposing the glue. While the hardened glue may not be very toxic, it isn't a recommended diet for goats or other animals.

Another material to avoid is paint with a lead base. If you

decide to paint your buildings, make sure the paint you use is nontoxic. Glass windows can be dangerous unless you put them high enough that goats won't be pushed into them.

When planning the barn, investigate prices of various materials to find the cheapest in your region. If you live near a small sawmill, you can probably buy rough-sawed, unplaned lumber at reasonable prices. Where no locally manufactured lumber is available, sheet metal may be the cheapest siding you can find. Seconds of this material are frequently offered by dealers at a substantial discount from the retail price. The defect is often no more than incomplete priming, which is a problem you can remedy.

Do you have a surplus of rocks on your property? If so, you can build a goat barn with little more than the rocks you pick up and some hand-mixed concrete.

FOUNDATIONS

Make the foundation solid and true so that your building will go up with a minimum of problems and will last. The foundation is the life of the building.

Many permanent buildings have concrete block foundations. For a large building such as a house, a footing trench is dug deep enough to extend below the frost line. Such a trench is normally about 16″ to 18″ wide, but for a small goat shed it need be no more than half that width. If you build such a footing, collect enough large rocks before you pour the concrete base. Place as many rocks as possible in the concrete without allowing them to touch each other.

The most common mistake in using concrete is to make the mix too wet. The corollary mistake is inadequate mixing. In choosing the proper amount of water to add, you must begin with too little and add just enough to make the mix plastic after thorough mixing. As a rule of thumb, don't use more than about 5 or 5½ gallons of water per sack of cement.

If you plan to dig the trench by hand and the soil is firm, you can use it as the form for the footing. Dig the trench carefully so that the sides are straight and clean. When the trench is complete, drive a line of stakes in the trench. Place these stakes along the center line of the trench. Use a carpenter's level with a straightedge, to make sure the tops of the stakes are level, about 2″ below ground level. Then pour the concrete until it is level with the top of the stakes (Fig. 4-3).

Unless you are planning an elaborate goat dairy, you can get by with a less extensive foundation. One kind of lower-cost foundation

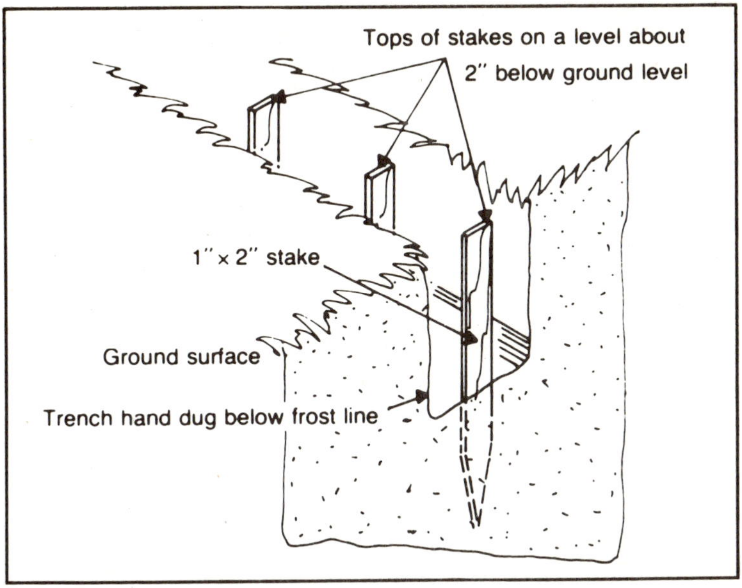

Fig. 4-3. How to use trench walls as a form when pouring a concrete footing. Pour the concrete to the top of each stake.

is to pour concrete posts or piers at intervals around the foundation. The building sills can be bolted directly onto these posts. Space the posts according to the kind of building material you will use. They can be spaced about 4′ apart for a typical small shed. To install the posts, dig a round hole, using a posthole digger, to a depth below the frost line in your area. Pour the concrete, using the hole walls as a form until you reach the surface level. Then insert a large can, which has had the bottom as well as the top removed, into the concrete. Forty-six-ounce juice cans are large enough for a small shed. Use a straightedge and level to make sure the top of each can shell is level with the others. Fill the can with concrete. While the concrete is wet, insert a 5/16″ bolt head into the center of the pier (Fig. 4-4A). The bolt should extend at least 2-½″ above the surface of the concrete. The bottom sill can be bolted to these piers.

In some parts of the United States many old houses were built with nothing more than wood posts for a foundation. This same technique can be used for a small goat shed (Fig. 4-4B). Wood post foundations are probably the most economical of any kind of satisfactory footing. The two most indestructible woods are *Osage orange,* also called bois d'arc, and *black locust,* but other kinds of wood can be used if treated with wood preservative. I know of one

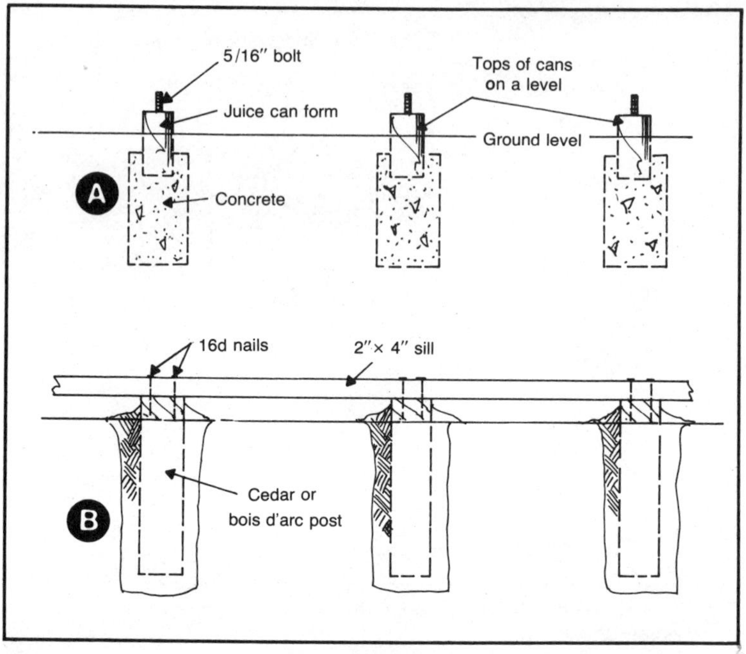

Fig. 4-4. Two ways to make a solid foundation for small sheds. (A) Poured concrete posts. (B) Durable wood posts.

goat shed that was built on bois d'arc posts about 25 years ago, and the posts are just as solid now as when they were new. The sills in that shed, three layers of oak 2" × 4"s, appear to be sound.

To install a wood post foundation, dig the postholes as you would for concrete posts. Set the posts, filling the space around each with a mixture of rocks and soil. Tamp the soil well around each post. Note that the posts need be only about 3′ long for this job if you are setting them 2-½′ deep. Don't use concrete to set wood posts. It will shrink away from the wood, leaving space for water to accumulate and encourage rot.

To level the tops of the posts, use a carpenter's level and the usual straightedge. Mark a line on each post with a pencil, then saw the post off at the line. If you cut a post too short, use a wood wedge to fill in the space between the first sill and the short post.

Another way to make a foundation is to use long poles treated with preservative (Fig. 4-5). The poles provide the vertical framing members as well as the foundation (Fig. 4-6). If you have Osage orange or black locust trees on your land, you may be able to build a barn from largely homegrown material. Remember that most other

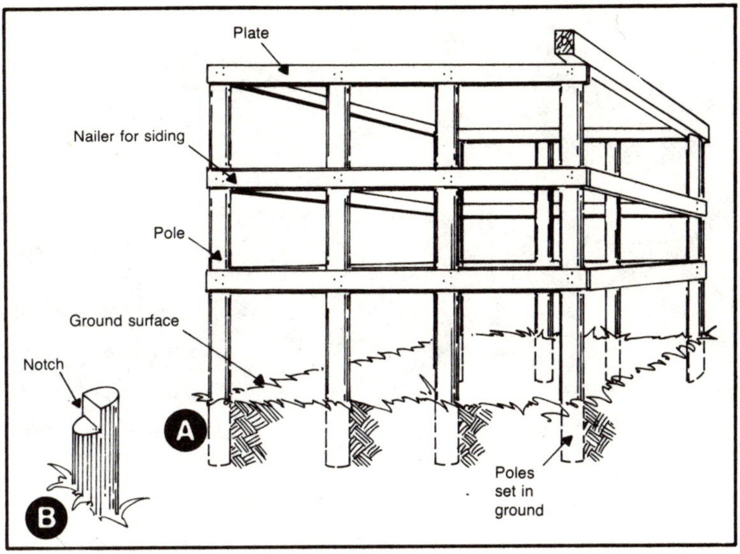

Fig. 4-5. Small shed with a pole frame. (A) Position of poles and nailers for siding. (B) Notch poles for plates and nailers.

woods must be pressure treated with preservative to last in the ground.

The shed shown in Fig. 4-6 was converted from a tumble-down, sagging shack into a useful shelter, suitable for a doe and supplies, by replacing the framing with cedar poles. The poles provided a combination frame and foundation for the building, which previously had only some loose rocks for a foundation. The man in Fig. 4-6 is adding a used 2″ × 4″ across the end poles to add strength to the original rafter than was installed incorrectly with the 4″ dimension horizontal. After installing cedar poles for stability and solidifying the existing structure with used lumber, a usable shed was available for a fraction of the cost of a new one. To give a light interior, the 36″ space above the horizontal 2″ × 4″ across the front poles was closed by installing plastic-coated screen wire, which was mounted on several hinged frames. The hinges permitted the frames to be opened for ventilation. You might try this kind of conversion if there is an abandoned shed in your neighborhood that you can buy cheap or obtain free for moving it. Such a building can make a satisfactory home for your goat.

FRAMING A SMALL SHED

The frame for a shed can be made with larger spacing between

Fig. 4-6. An old shed can be rebuilt, using a pole frame, to make an acceptable low-cost goat shelter.

studs than that of a larger building. The typical goat shed will be no more than about 6-½″ tall at the eaves, so the weight load on the frame is low. In earlier times most farm structures had a loft for hay storage, which required a better frame. A loft for hay requires more work in raising the heavy bales up to the loft, then getting them down again. The chief advantage is that less roof is needed than for a single story building of equal floor area. A two-story building requires a stronger frame and also a floor for the loft. If ground space is severely limited, a loft may make sense.

If you build the goat shed yourself, you can use a lower grade of framing lumber and plan to use the individual beams so that weak places, such as large knots, can be cut out or reinforced. Avoid putting these flaws in a horizontal beam at the midpoint between support points. Sort your lumber and select the best pieces for sills and plates. Sills for small sheds can be made from 2″ × 4″s. If necessary, use a double layer of 2″ × 4″s to make a 4″ × 4″ sill (Fig. 4-7).

As mentioned earlier, you can make the center to center distance between studs greater than the usual 16″ used in large buildings. Don't skimp on rafters, especially if you live in snow country. It is better to use 32″ distances between wall studs than to increase the distance between rafters. The siding will add strength to the walls, but sheathing adds little resistance to the pull of gravity on a roof.

Choose a roof pitch to match your climate. Sheds in regions

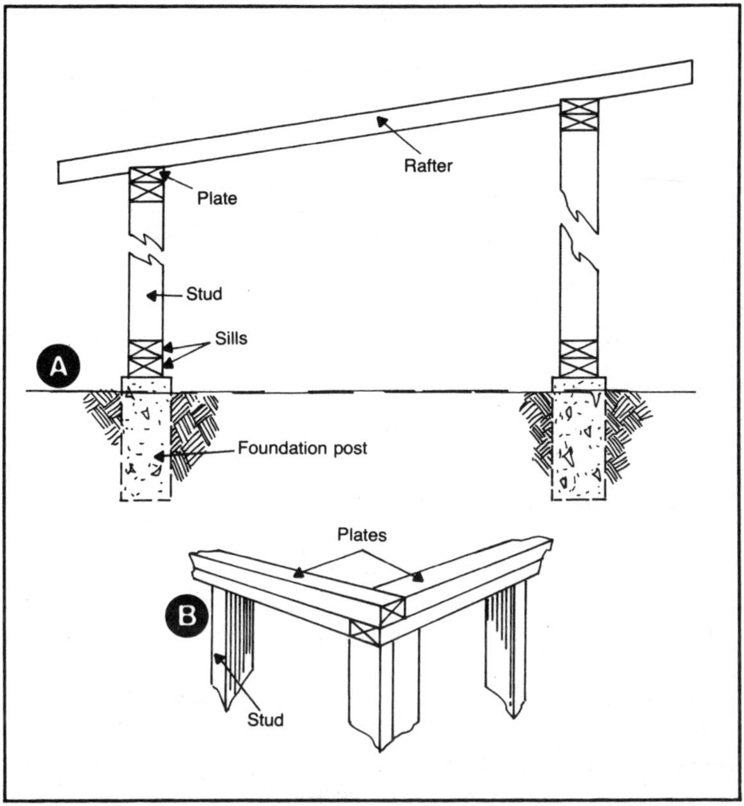

Fig. 4-7. Suggested framing details for small shed. (A) Position of sills, studs, plates, and rafters. (B) Use lap joints at corners.

with heavy snowfall require more pitch than warm climates. As a minimum in a dry region, the pitch should be at least 1:6, which would place the higher plate in a building with a 10′ span 2′ above the level of the lower one.

DEVELOPING A FLOOR PLAN

If you plan your barn before starting work, you can eliminate much of the daily labor required by goats in a makeshift shelter. One great labor saver, borrowed from 4-H goat leader Ruth Scott, is to place the hay storage next to the hay feeder (Fig. 4-8). To further reduce labor, try to make the hay storage area accessible to vehicles for easy hay unloading. Make the door to that section larger than the typical shed doors. All doors should be large enough to get a wheelbarrow in for easy cleaning (Fig. 4-9).

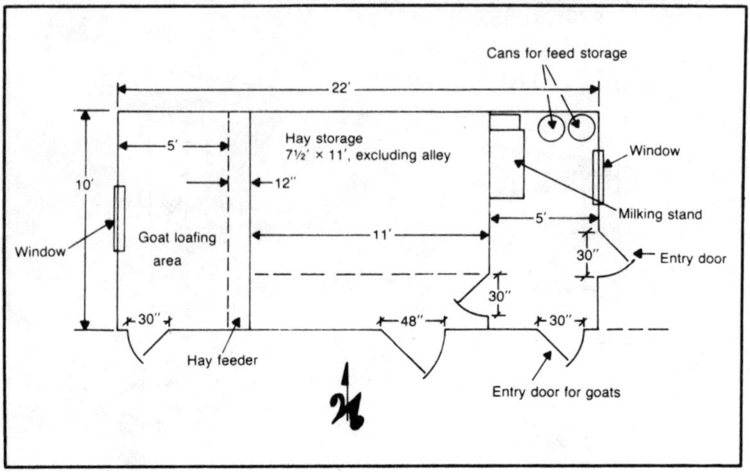

Fig. 4-8. Floor plan for a small goat barn.

How to Plan for Adequate Hay Storage

Calculating the necessary hay storage area is one of the primary steps in planning a shelter. If you fail to provide enough space, you will have to constantly replenish your supply, probably at times when hauling hay is inconvenient and the price is high. With sufficient storage space you can often buy your hay in the field and haul it yourself, with considerable savings.

Refer to Table 4-1 to get a rough estimate of the amount of floor area needed for hay. Select the column that corresponds to the height from floor to ceiling in the hay storage area of your barn. When you find the number of square feet of floor area needed per ton, multiply that figure by the number of tons you require. If you have a 10′ ceiling and need storage space for 4 tons, you would find the number of square feet per ton, 27, and multiply it by 4 to get 108 square feet.

If you live in the far west, you may find that much larger bales of alfalfa hay are the normal size. These large bales, made with three ties per bale, usually weigh about 300 pounds and are handled by mechanized equipment. In many regions the trend is toward larger round bales that are stored outdoors. Find out what is available in your area and the necessary gate or door widths needed to accommodate the kind of vehicles that will deliver your hay before you make your building plans.

Here is how the barn space was allotted for the plan in Fig. 4-8. It was assumed that the barn would be adequate for two does. The annual hay requirement for a doe varies from animal to animal and on the availability of pasture. Assuming no pasture, a doe may require from 1000 pounds to 2000 pounds (1 ton) of hay a year. Taking the higher figure, provide a hay storage space that can hold 2 tons of hay. The amount of hay in a rectangular bale varies greatly according to how the baler is operated, but with proper baling it should equal about 60 pounds. About 67 bales would be required to equal 2 tons. A normal bale size is about 15″ × 18″ × 42″. If you make the height of the building, at the low end, 6-½′ or 78″, the bales can be stacked five layers high. Dividing 67 by 5, then rounding off, you can see that each layer must contain 14 bales. Five layers of 14 bales each provides storage for 70 bales, more than enough for two does under any normal conditions.

To get the area required for 14 bales in each layer, you can use various combinations of bales such as 3 × 5, which allows one bale extra, or 2 × 7. The 2 × 7 arrangement was chosen because it fit in well with the available space in the original building plan. The area occupied by two rows with seven bales in each row is 84″ × 126″, or 7′ × 10½′. In the final plan a slightly larger area, 7-½′ × 11′, was provided because it was available. You should never skimp on hay storage. Some bales will be irregular or bulgy, but an extra 6″ each way permits space to accommodate those bales.

In regions that have unusually cold winters with many days of subzero weather, it may be desirable to use hay as insulation.

Fig. 4-9. A goat barn based on the floor plan shown in Fig. 4-8.

Table 4-1. Floor Area Required to Store 1 Ton of Hay in Barns of Various Ceiling Heights.

Ceiling Height (ft.)	6-1/2	8	10	12
Approximate Floor Area (sq. ft.)	37	32	27	21

Design the hay storage area so that it wraps around the goat loafing space on three sides—east, west, and north. Provide ventilation if this is done. For a building with a simple shed roof, the vent area can be located just below the roof at the front, or high end, of the shed.

Loose Housing Goat Space

After providing necessary hay storage, the next step is to make sure there is enough loafing space for the goats. Here you are planning a loose housing area. This space is perhaps even more important than hay storage because you can get by with minimal hay storage, but the goats obviously must have shelter from cold winds and dampness. The hay storge was considered first here because it is too often overlooked until the plan is almost complete, and the hay gets only the surplus space.

Each goat needs at least 20 to 25 square feet of loose space in the "family room." You need about 40 to 50 square feet for two does. The building plan shown in Fig. 4-8 permitted a width of 10′, a figure chosen so that rafters could be made from 12′ beams. Although a space 4′ × 10′ might be adequate, provide 10 more square feet by making the goat area 5′ × 10′. That will permit temporary space for small kids.

The plan shows no special provision for kidding rooms, but a folding gate can be temporarily installed across the goat area to divide it into individual 5′ × 5′ pens. By keeping breeding records on your does, you will know when a doe is due to kid. You can install the gate to give her privacy during that time. The kids may be kept with their dams for a short time, then moved to their own pen. Some of the hay storage area can be used temporarily for kids. Kids will normally arrive in late winter or early spring when the hay supply is somewhat depleted.

Milking Room

The last part of the plan to be developed is the milking room.

The area shown, 5′ × 10′, is more than enough for a milking stand. This area can be used as grain storage and for general storage of tools and supplies. Any surplus storage cabinets can be used here. Garage sales often can provide these items at a cost below that of the materials needed to make them.

The plan shows two windows, one for the milking room and one for the goat space. Almost any kind of windows can be used, so be alert to bargain items from yard sales. A storm window is an excellent choice of an all-metal window that can often be picked up at a low price. Make sure the windows are high enough that the goats won't be likely to break the glass. If you can't find a bargain glass window, you can leave an open cutout and make a solid door to close during bad weather. You can wire the shed for light at night. Without electrical wiring you will often be forced to use flashlights to tend the goats, especially in winter when the days are short. Don't resort to candles.

Don't assume that Fig. 4-8 represents the best of all worlds for your goats. It was presented as a guide to how one plan was developed. You may get some hints on how to design a barn for your special situation by following through on the design process. A goat barn can be little more than a lean-to, or it can be a highly professional structure. A milking parlor with concrete milk stands and running water is necessary for commercial production, but it is not very practical for the backyard dairy.

HAY FEEDERS

When hay is fed in open racks, goats waste much of it because they usually refuse hay that has dropped to the barn floor. This hay can be contaminated with parasites. To avoid waste and insure that parasitism is kept to a minimum, the hay feeders should be designed for goats.

The typical solution is to install keyhole feeders (Fig. 4-10). The goat can insert her head at the normal level and then lower it to the feeder floor, which should be from 8″ to 12″ above the barn floor, to get the hay. She can't withdraw her head at that level. This forces her to either keep her head in the feeder while eating or to raise her head before she withdraws it. When she raises her head, much of the excess hay that dangles from her mouth falls back into the feeder before she can remove her head through the keyhole. Hay that would be wasted on the barn floor thus stays in the feeder.

Some goat keepers feel that the points at the upper part of the cutouts on the keyhole feeder can injure goats and perhaps cause

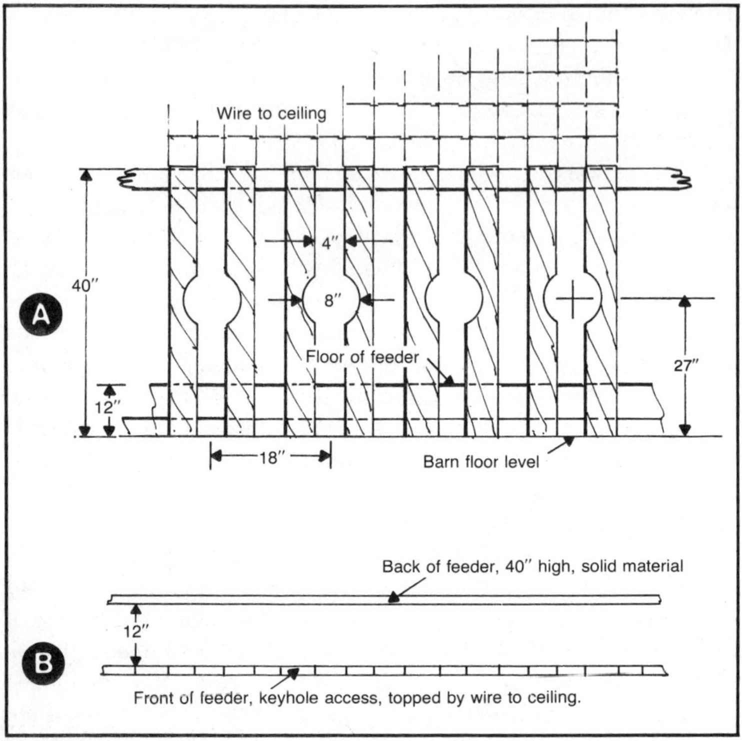

Fig. 4-10. Suggested plan for a hay feeder. With the wire partition above the feeder, hay can be stored behind the feeder for easy feeding. (A) Front view of the hay feeder. (B) Top view.

abscesses. They recommend that the keyhole be lengthened to make slots that are wide enough at the upper end for the goat's head, but not at the lower end (Fig. 4-11). This style is mandatory for most goats with horns.

The dimensions for such a feeder are not critical and can be adjusted to fit the size of your goat. The keyholes should ideally be placed about 18″ apart (Fig. 4-10), but you can place them much closer if you are crowded for space at the feeder (Fig. 4-12).

In addition to saving hay, keyhole feeders eliminate much of the fighting that occurs when hay is put out in open racks. Hay should never be fed from the ground (Fig. 4-13) unless it is on new fallen snow or in an area not usually populated by the goats.

One goat keeper uses his hay storage space as a no-labor feeder. He installs stock panels (more about these later) as partitions between the hay storage area and the goat space. Goats can

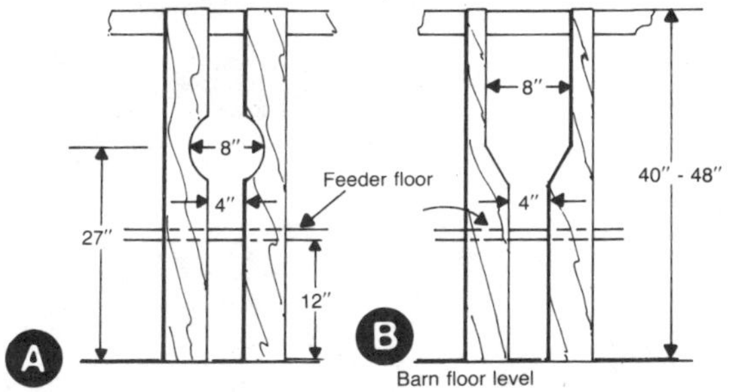

Most dairymen use the keyhole feeder (A), but for horned goats the alternate feeder (B) is most desirable. some goat keepers report injuries with the keyhole feeder because of the narrow opening at the top of the keyhole.

Fig. 4-11. Suggested dimensions for hay feeder access holes. (A) Keyhole feeder. (B) Enlarged slot feeder.

reach through the space between the bars in the stock panels to get hay (Fig. 4-14).

GRAIN FEEDERS

Much of the grain feeding of does will be done at the milking stand. Where the does are kept in separate stalls, grain feeding can

Fig. 4-12. A hay feeder with keyholes at various heights to accommodate goats of different sizes.

Fig. 4-13. Don't use everything, including the kitchen sink as was done here, to make a hay feeder. The goats will waste much of the hay that is put out.

Fig. 4-14. An easy-to-make self-feeder. By storing bales of hay behind a stock panel, the goats can reach through the panel and eat from the bales.

be done by installing a feed box in each stall. Individual grain feeding is easy where tie stall housing is used, but a communal feeder is the rule where goats are kept in loose housing. You can furnish individual feeding in loose housing by providing some way to separate the does at feed time. If you have no more than a couple of does, use a gate to separate the does while they are eating. In warm climates you can feed one doe outside the shed while the other eats inside. Whatever you do, set up a regular habit of feeding the does in the same sequence and location every day. They will soon adjust to your routine.

You can build a grain feeder for common feeding of several animals that is much like a small hay feeder. An enlarged version of the kid feeder (Fig. 4-15) can be used for does. The only disadvantage is that the fast eaters may get more of the grain. One advantage is the competition often spurs finicky eaters to eat better, especially if you are feeding a ration they don't find appealing.

GRAIN STORAGE

Most grain dealers now sell grain or mixed feeds in 50-pound paper bags. Large operators typically leave the grain in bags until

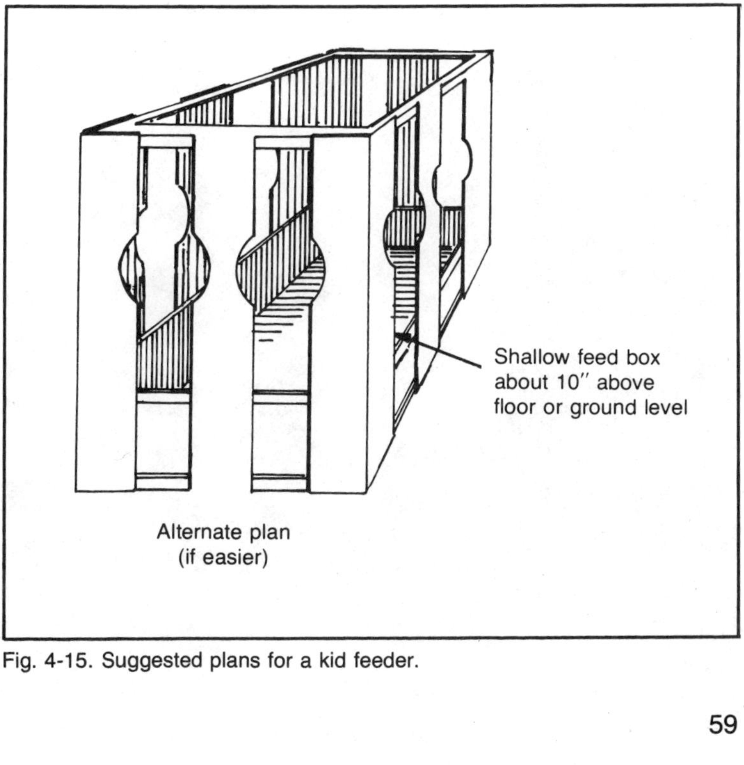

Fig. 4-15. Suggested plans for a kid feeder.

use, but paper can absorb moisture in damp climates. Regardless of your climate, rodents may be a problem with stored grain. The simplest solution is to get one or two large 30-gallon garbage cans that have tight lids and store the grain in them (Fig. 4-16). Don't buy grain too far ahead of your planned use for it, especially in hot weather. Goats don't like moldy grain. It can cause health problems in young or pregnant animals.

MILKING STANDS

A milk stand puts the goat up to a level where you can milk her without squatting on the ground (Fig. 4-17). If you are tall, you may

Fig. 4-16. Protect feeds from contamination by using closed containers in a location where goats can't get to them.

Fig. 4-17. A crude milk stand will put the goat up to a level where you can milk without squatting on the ground.

want to build an extension to the stand—a place for you to sit. You can use a portable stool, but a stand is almost essential.

The stand can be as primitive as the one in Fig. 4-17, or it can include a stanchion with provision for locking the goat's head. Another way to hold restless goats in place is to use a chain and snap that connects to a ring in the goat's collar.

To build a milking stand, start by making a frame (Fig. 4-18A). Almost any surplus material can be used, but the basic frame should be built from 2″ × 4″ or 2″ × 6″ lumber. Note that eight pieces are needed for the frame shown in Fig. 4-18A. You will need four pieces 18″ long, two pieces 36″ long, and two pieces 48″ long. Select these pieces from good strong lumber with no weak points. Assemble the frame with 16d nails, then cover the platform area with 1″ boards. For an easy-to-clean surface, get some surplus linoleum and cover the platform with it. Build a small feed box and install it on the uprights at a reasonable level for your goat (Fig. 4-18B). Check the finished stand for rigidity. If necessary, add diagonal braces to the legs (Fig. 4-18).

Some plans for goat stands show a ramp leading to the platform for the goat to walk up. This is rarely necessary because most does

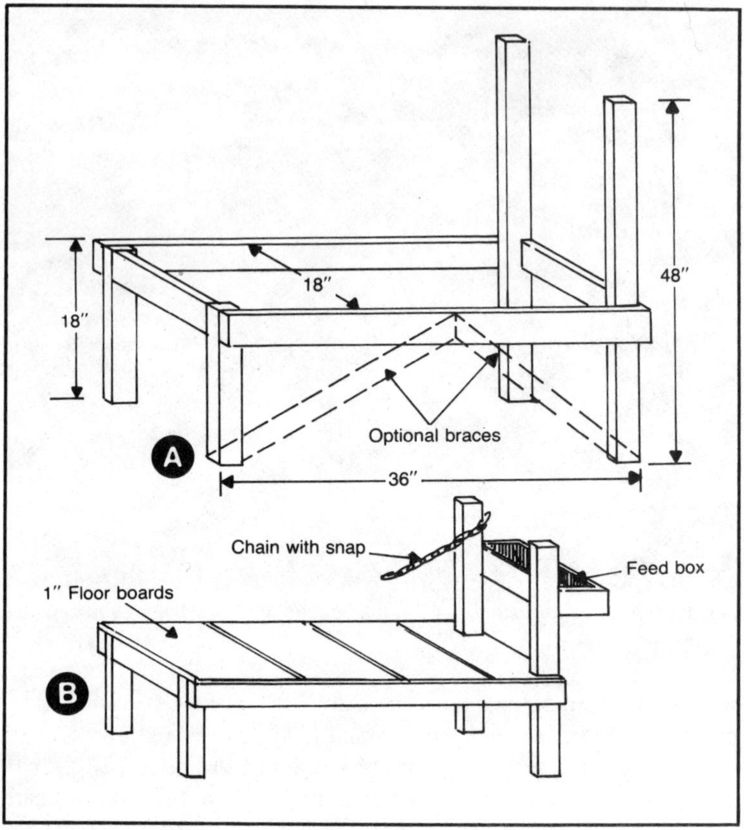

Fig. 4-18. Suggested plans for a milking stand.

are agile and can easily hop onto a milking stand without danger of injury. If you have a goat with a pendulous udder that could be injured by the doe jumping onto the stand, add an extra step or ramp at the rear of the platform.

BUNKS

If given a choice, goats will usually sleep on an elevated surface. Some goat keepers build bunks for their goats. For a barn with limited space the bunks can be made in tiers, one above the other. The dominant goat will usually choose the highest level, the second in order will take the next level and on down to the bottom goat. Three levels, with the platforms about 18″ apart, are probably as high as you should go with this arrangement. Make the bunks about 2′ wide and long enough to accommodate one goat. You can sleep more than one goat on a level by making the bunks longer. Some goat owners prefer to use only a single level close to the floor. One man explained his reason, "I don't want my goats packed in like the sardine can troopships of World War II."

KID PENS AND HOUSING

The number of kid pens and separate housing quarters that you will need depends on the kind of goats you intend to raise. If you are keeping grade animals, you will castrate the young males and can use the same pen for doelings and wethers. If you are raising registered stock, you will be saving at least some of the better male kids for sale as bucks. They will need a separate pen and shelter to avoid unplanned breeding of doelings. Don't plan to keep your baby buck kids in pens with mature bucks; the older bucks can easily injure the kids unintentionally. You may be tempted to use a common pen as temporary housing for buck kids and doelings, but remember that young bucks are precocious. This kind of temporary solution can lead to a more permanent problem of a doeling stunted by too early pregnancy.

Plan your kid housing to avoid the buildup of parasites. Some goat keepers prefer to put their young kids on an elevated floor made from heavy hardware cloth or wood slats. These floors permit the droppings to go through instead of collecting in the bedding. The bedding should be some material that kids won't eat. Baby goats will chew on almost anything that resembles food such as straw or moldy hay. Sawdust or wood shavings make good bedding for kids, and they aren't likely to eat such materials.

Feeders for kids should be made smaller than those for adults (Fig. 4-15). The feeder shown in Fig.4-15 was designed for use in the pen outside the kid barn. Indoor feeders should be placed so that the kids have access to them only through the usual keyhole partitions. A kid house can be built on a plan similar to that of the hay storage and loafing sections of Fig. 4-8, but with reduced dimensions.

Kids enjoy an elevated platform on which to play (Fig. 4-19). An exercise board can be provided by nothing more than a plank supported by short posts. The one shown in Fig. 4-19 was converted from an old hay feeder.

BUCK SHEDS

A buck should be protected from rain, snow, and cold winds. A buck shed can be just large enough to comfortably house him. Some

Fig. 4-19. Kids enjoy any elevated surface, such as this converted old hay feeder, on which to play.

goat keepers have successfully adapted large doghouses for bucks. The buck must be fed outside with such a small shelter, but that is no problem in most climates. Because of the buck's reduced housing needs, you can build a buck shed from almost any kind of salvage material (Fig. 4-20).

One woman who keeps goats has made instant goat houses from the shells of some old car bodies. This is an acceptable solution to the housing problem only if you don't mind having your place look like a salvage yard, and you have a suitable screen to hide the derelicts. It isn't a useful option if you have neighbors who would object to a car body rusting out in your back lot. This alternative is probably illegal in most localities unless a tall fence or other screen hides it from the street or highway.

The buck shed shown in Fig. 4-20 is an example of a minimum-cost shed that can offer adequate protection for a buck or any goat. It was built in an hour or so by the owner. He got some surplus pallets from the discard pile at a furniture factory. Then he went to a newspaper office and bought several sheets of thin aluminum at 40 cents a sheet. It took only a minute to set up the pallets and nail them together, supported by a post. He covered the open framework of the pallets with four or five sheets of aluminum to complete the shelter. It has been used for two years in all kinds of weather,

Fig. 4-20. This small buck shelter was built in about an hour using surplus pallets and aluminum printing sheets from a newspaper office.

including subzero temperatures, and the goats housed there have remained healthy. If you are low on cash, look around for cheap surplus materials. Use a little imagination and install temporary goat shelters that will serve until you have the money and time for a permanent setup.

Chapter 5

Fencing for Goats

Good fences can increase your efficiency in every aspect of goat keeping. Fences save time and labor in managing the goats for proper feeding, breeding, and disease control. They also enable you to make use of your ground for pens or pasture. A good fence will protect your trees, shrubs, or other plants from the goats (Fig.5-1). If you have close neighbors, fences can prevent difficult situations that occur when livestock strays from home.

Some goat keepers make do with fencing that others find inadequate. I once visited a woman who keeps goats in her backyard in a small city. When I saw the goats placidly chewing their cuds behind an unusually low fence, I asked if they ever got out. She said that they had never jumped the fence. There was no fence charger. I admitted that my own goats wouldn't stay in such a lot.

"You have brush goats," she said.

I ignored the insult, knowing that my goats were as good as hers and were of various breeds. When I returned one night to find my goats lined up for roll call as the car's headlights swept the front porch, I wondered again why they escaped at the slightest opportunity while the city goats stayed in their place. My fences weren't perfect, but they were far better than the ones I had seen at the town place. Someone suggested that urban goats may be more civilized than rural goats. I finally learned there was some truth to that remark. I also learned what to do about it.

Fig. 5-1. Stock panels make a superior fence for goats but are expensive.

EDUCATIONAL PSYCHOLOGY FOR GOATS

The problem of choosing a good fence for a goat has two parts to it—the fence and the goat. I got a clue to the behavior of my goats only after considerable experience. It had nothing to do with their genes.

I noticed that even when I got a new goat that had no history of breaching, it soon jumped fences at every chance. Every year I traded bucks and other goats with a man who lived in a remote hill region. He allowed his goats to roam over some wooded hills that were crisscrossed with the remains of old fences. The goats grew up jumping barriers, from the trunks of fallen trees to knee-high remnants of fences. When the goats reached adulthood, they could jump very high.

One year I obtained two wethers for butchering from the same man and put them in a lot with a 4′ fence. They immediately leaped over the fence. I didn't want to tie them on short leashes because that would surely excite them. A friend suggested that I tie them together with a short rope. They could then roam the lot together, but if one tried to jump the fence, they would be caught. Because neither one could make a run at the fence, jumping would be impossible. That was the theory. I tied them together with a piece of rope about 2′ long. When I looked for the goats a few minutes later, they were trotting across an adjoining field, still tied together like a team of horses.

That incident helped me solve the problem of jumping goats. It proved that the goats from the hill country were conditioned to jumping any obstacle. When I put them into my herd, all the goats quickly adopted the bad habit. Start training your goats to respect fences while they are young.

The most important step in the training program is to never put a goat in a pen that is easily breached. Always move young goats through gates (Fig. 5-2) never lift them over a fence (Fig. 5-3). Once a goat gets the habit, it is almost impossible to stop it.

One method of altering a confirmed jumper's behavior is to keep the animal in a pen with a woven wire cover. After the goat learns that she can't get out by jumping, the habit may weaken. Maintain your fences at all times to keep her in the pen. In no other endeavor does the old adage about a stitch in time apply more aptly than in fencing goats. In my early goat raising days I tried to solve the problem by escalation. I would build a 3′ fence, and the goats would jump it. I would add another 6″ of fencing. The goats would jump that fence. It was like raising the bar at the high jump event in a track meet. Every athlete knows that you can jump higher if the bar is raised gradually than if you must clear the highest level on your first jump. The first fence must be adequate for goats.

KINDS OF FENCING

More goat fences are probably made from *woven wire,* the kind designed for hogs and other livestock, than from any other kind (Fig. 5-4). This fencing is good for almost all animals, but it is less than ideal for goats. Goats like to stand with their front hooves on the

Fig. 5-2. When moving kids, always take them through gates.

cross stays, causing the fence to sag (Fig. 5-5). Another problem with most woven wire fence is that a horned goat will force her head through to reach an irresistible bite of browse on the other side, then she is caught fast (Fig. 5-6). If you use woven wire fence, there are several ways to avoid the problem of caught goats. The best way is to dehorn the kids. Another way is to ask for sheep fence that has a 12″ distance between vertical stays instead of the more common 6″ distance. You can also install an electrically charged wire in front of the woven wire fence to keep the goats away from it.

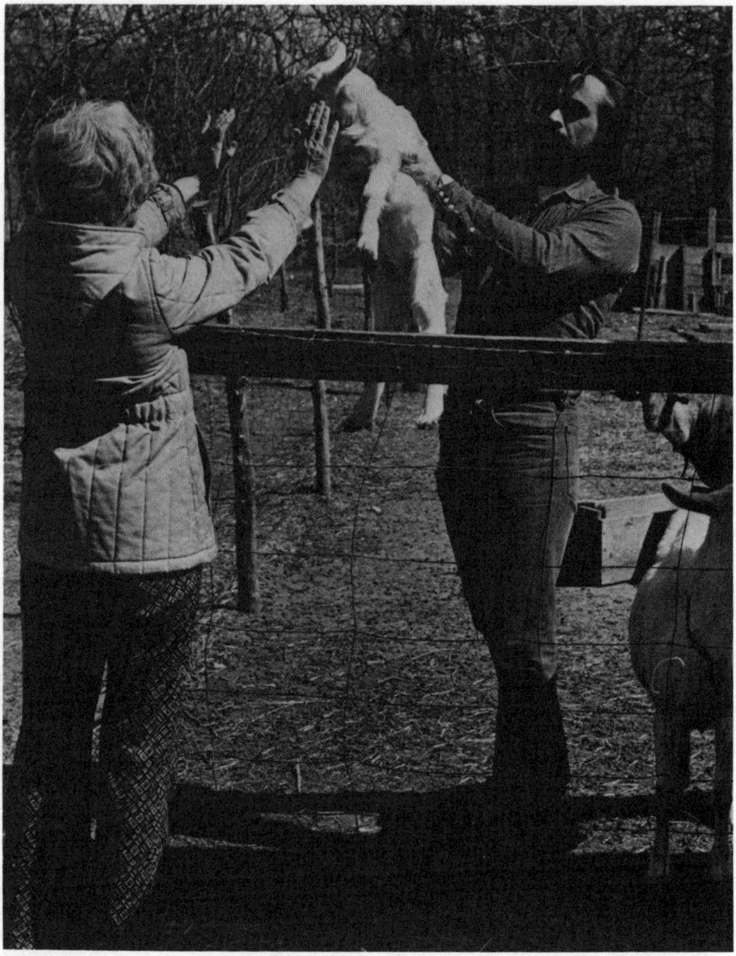

Fig. 5-3. Never lift kids over fences.

Barbed wire fences have long been popular with most farmers and ranchers, but they are dangerous for goats. There are no vertical stays, so goats often try to go through the fence. They usually injure their teats and udders. One possible use is to extend the height of an existing fence by stretching one or two strands of barbed wire above it (Fig. 5-5). The possibility of udder damage is reduced by using barbed wire only at the upper part of the fence.

Board fences are sometimes used for pens, but they are probably too expensive for larger lots or pastures. Avoid fences with sharp points at the top.

One kind of satisfactory fence is the *close mesh welded wire fencing,* sometimes called stock fence (Fig. 5-7). The typical mesh for goats is 2″ × 4″ spacing. The vertical stays are 2″ apart, and the horizontal stays are 4″ from each other (Fig. 5-8). There is no possibility that a goat can reach through the fence with openings no greater than 2″ × 4″. The chief disadvantage of this fence is that it is about three times more expensive than woven wire. Stock fence is available in 100′ rolls, while woven wire is sold in 330′ (20 rod) rolls. Remember this difference when comparing prices.

Fig. 5-4. Most goat fences are made with woven wire.

Fig. 5-5. Goats like to stand with their front hooves on fences, causing the fence to sag.

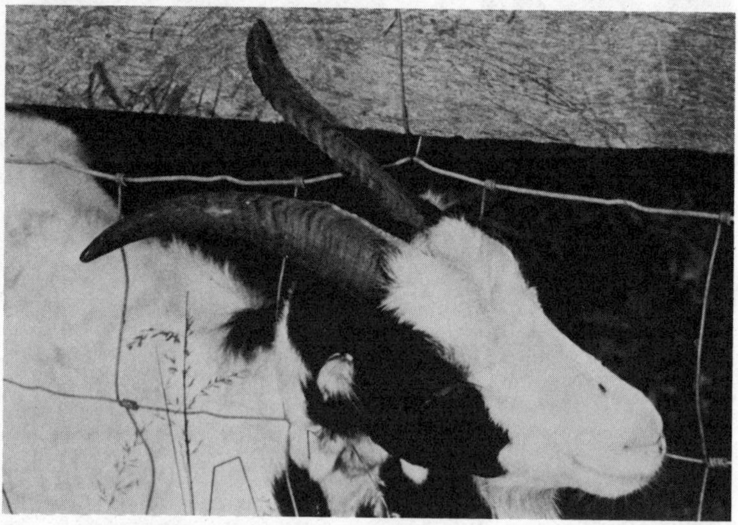

Fig. 5-6. Goats with horns often get caught in woven wire fencing that has the vertical stays at 6″ intervals.

Fig. 5-7. Here stock fencing, reinforced by boards, is used to protect the trunk of a tree in a goat lot.

Stock panels are perhaps the best goat fencing (Fig.5-1). Stock panels are heavier than ordinary fencing material, because they are made by welding ¼″ rods together. These panels are too stiff to be sold in rolls. They come in 16′ lengths and in 34″ and 52″ heights. The lower panels are for hogs, but they are too low for goat pastures. Get the cattle panels. Farmers use these panels to make temporary pens, gates, or partitions in barns. The goats cannot bend or damage them by standing on the stays. Because the panels are so rigid, they require fewer posts. Stock panels are too expensive for general fence use.

Which kind of fencing material is best? You must weigh cost against performance. Even though it has serious disadvantages, woven wire is chosen by most goat keepers because it is cheaper than welded wire. If you choose woven wire, select a heavy gauge. It usually costs little more than light gauge fencing and lasts much longer.

KINDS OF POSTS

The three kinds of post material in general use are wood, steel, and concrete. At one time most farm fences had wood posts, but steel has gained in popularity. Although steel usually costs more, it

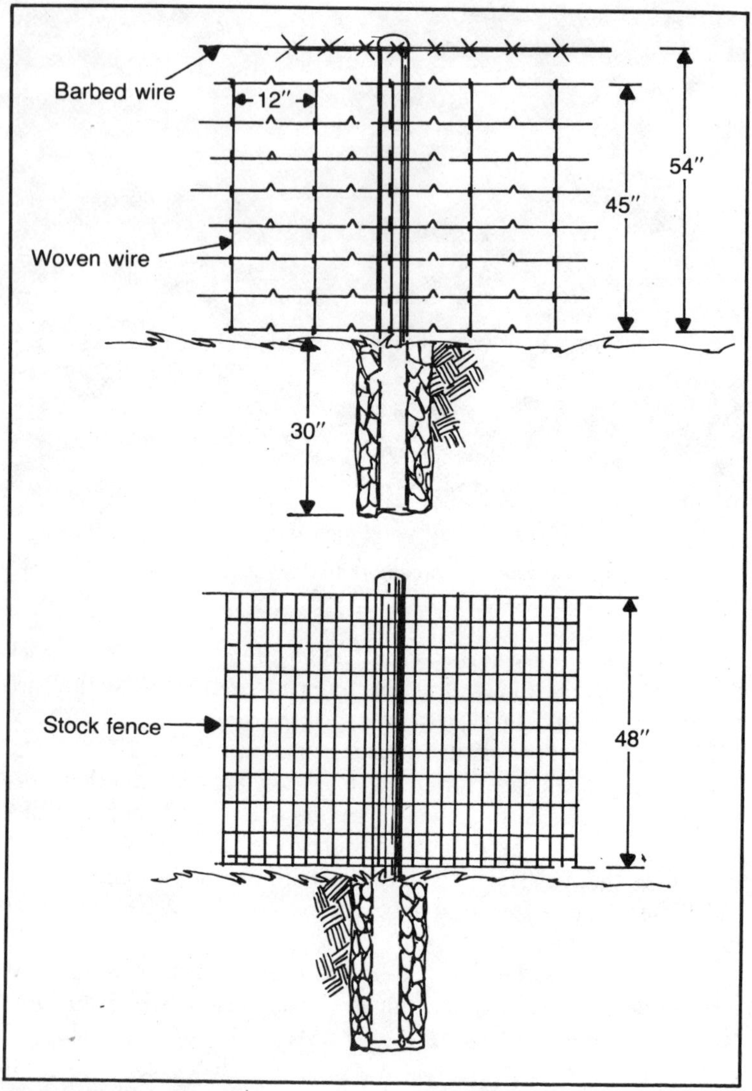

Fig. 5-8. Typical dimensions for woven wire and stock fences.

has several advantages. One advantage for fences on rocky soil is that the thinner steel posts can be more easily driven into the ground, saving labor. They help prevent death to animals from lightning by providing some degree of grounding for the fence. You can ground any fence by driving a long galvanized rod into the ground and connecting it to the wire fence with a #6 gauge copper wire. The galvanized rod should be at least 6′ to 10′ long. Such ground rods should be inserted every few hundred feet along the fence.

If you choose steel posts, you may want to use large wood posts for corner posts. Heavy gauge angle iron, 2-½″ × 2-½″ with a thickness of ¼″, is also good. You can use other kinds of scrap iron bars, such as a section of rail from a railroad, if you have them or can buy them at a reasonable price. In selecting wood corner posts, get a treated post that is at least 6″ in diameter or larger. The life of the corner post is the life of the fence.

Wood posts may cost less than others, but beware of untreated wood posts unless they are made from black locust or Osage orange trees. Osage orange posts last at least 25 years, but the posts are usually crooked and, as the wood seasons, they often crack and release staples. These are not fatal flaws. You can set a crooked post so that the crook is in line with the fence rather than across the line, and you can use pieces of wire to tie up any points where staples have fallen out. Osage orange is more difficult to use than most posts. Large Osage orange posts that are acceptably straight make good corner posts.

Black locust wood lasts about as well as Osage orange (Table 5-1). Cedar is next in quality, with an expected life of 15 years or more. Redwood ranks just below cedar in post longevity.

If you buy your posts from a lumberyard or post company that supplies posts treated with a preservative, many kinds of wood are acceptable. Try to learn how the posts are treated. The reputation of the post company will usually tell you what to expect in quality

Table 5-1. Approximate Length of Life for Various Posts.

Untreated Posts	Treated Posts	
25 Years:	25 Years:	
Black Locust	Bald Cypress	Red Cedar
Osage Orange	Catalpa	Sassafras
	Cedar	Sweet Gum
	Hemlock	Sycamore
	Pine	Yellow Poplar
15 Years:	20 Years:	
Bald Cypress	Aspen	Maple
Catalpa	Basswood	Spruce
Cedar	Birch	Sweetbay
Red Cedar	Box Elder	Tamarack
Red Mulberry	Butternut	Tupello
Sassafras	Hickory	Willow
White Oak	Honey Locust	White Oak

control. Some post sellers offer treated posts that have not been subjected to enough pressure during the treatment process. Posts from these sources will rot in a few years, starting at the center where the treating solution failed to penetrate.

Concrete posts are durable but heavy to move and install. Another disadvantage is that you must provide some method of attaching wire to the concrete. One way is to make holes in the posts by inserting oiled rods through the forms. These posts typically have a square cross section of about 4″ × 4″. Many fence builders have sworn off concrete because of failure with it. One reason is lack of reinforcing iron. Two double lengths of ¼″ rod are suggested, each bent double and tied together with wire. Place the reinforcing rods so that they are evenly spaced near the four corners of the post, but covered with ¾″ to 1″ of concrete. Tamp the concrete well after pouring. Allow it to cure, the longer the better, before using. The concrete should ideally be kept damp for weeks or months to retard fast setting. The longer it takes to cure, the stronger the posts.

Because of the problems with concrete, your choice is likely to be steel or wood. If you live in a rural area, you can sometimes buy used steel posts at a saving.

SETTING POSTS

To give your fence a professional look, try to set the posts in a straight line, at the same height, and with equal spacing. Set the end or corner posts first. Then stretch a wire between them as a guide for the line posts. Although posts for woven wire fences are usually set 16′ apart, you will have a stronger fence if you use 12′ spacing. To get the posts set at equal distance from each other, make a gauge pole as long as the distance you plan to use. Use the gauge pole as a spacer when digging holes or driving posts. If you are driving in steel posts, wrap a piece of tape at the same point on each post to mark the ground surface level. If you are using 7′ posts and you expect to drive them 2-½′ into the ground, you would apply the tape at a point 2-½′ from the bottom of each post. If you are setting posts, place a piece of tape at the proper point on your posthole digger to get all the holes at the same depth.

Setting posts will require more labor and time than driving sharpened posts, but it is recommended for wood posts. Sharpening a wood post removes wood at the bottom where the post is susceptible to rot. You will need an auger or a posthole digger to dig holes by hand. If you want to complete the fence in a hurry, check if anyone in your neighborhood does this job with mechanized equipment. A

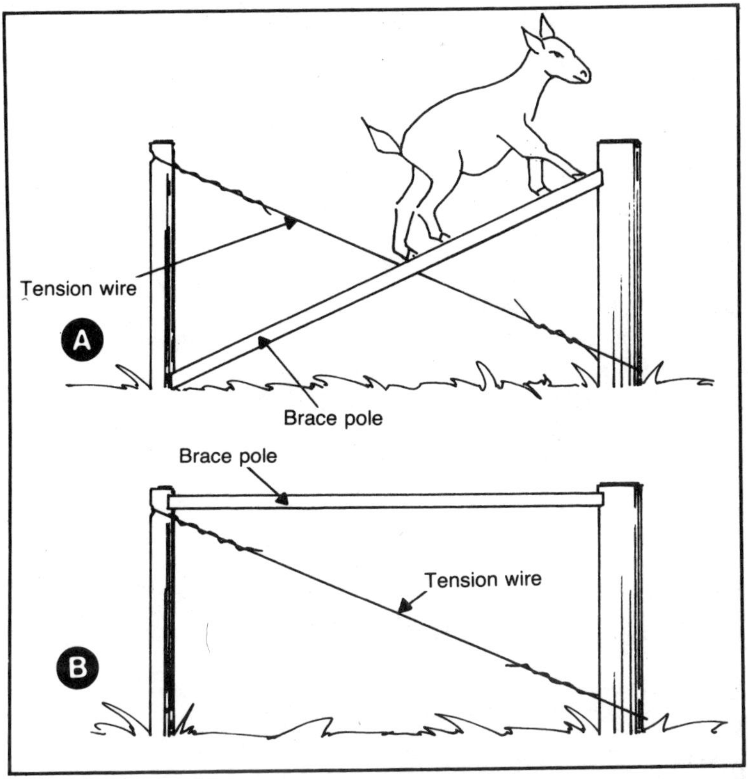

Fig. 5-9. How to brace corner posts. (A) Method used for many stock fences. (B) A better choice for goats. There is no sloping brace pole for goats to climb.

tractor operator with a posthole digger attachment can dig the holes about 20 times faster than you can by hand.

Before stretching the wire, make sure the end or corner posts are properly braced. Many farmers routinely use a sloping brace pole when building fence (Fig. 5-9A). This is not a good choice for goats, because kids and some adults will climb the sloping brace and then jump the fence. Use a horizontal brace pole to avoid this problem (Fig. 5-9B). Use the following procedure to install a brace pole:

☐ Notch the posts to fit the brace pole.

☐ Drill ⅜″ holes in the pole and posts to receive steel pins (Fig. 5-10).

☐ Wrap the ends of the brace pole with several strands of wire and twist tight.

☐ Drive steel pins through posts into the ends of the brace pole.

Install tension wires to tighten the braces. Use the following procedure:

☐ Notch the posts a few inches below their tops to receive tension wire.

☐ Wind four strands (two loops) of #9 smooth galvanized wire around posts (Fig. 5-11A).

☐ Loosely staple wire, if necessary, to hold in place.

☐ Insert a short stick between the loops of wire and twist until tight (Fig. 5-11B).

☐ Leave the stick in the wire for future tightening, if needed.

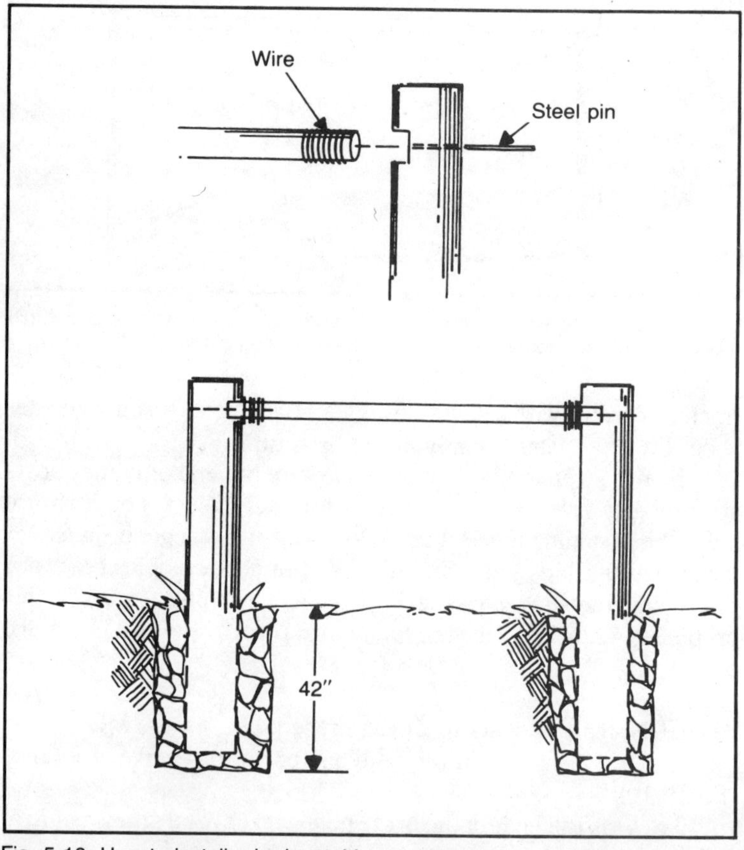

Fig. 5-10. How to install a horizontal brace pole.

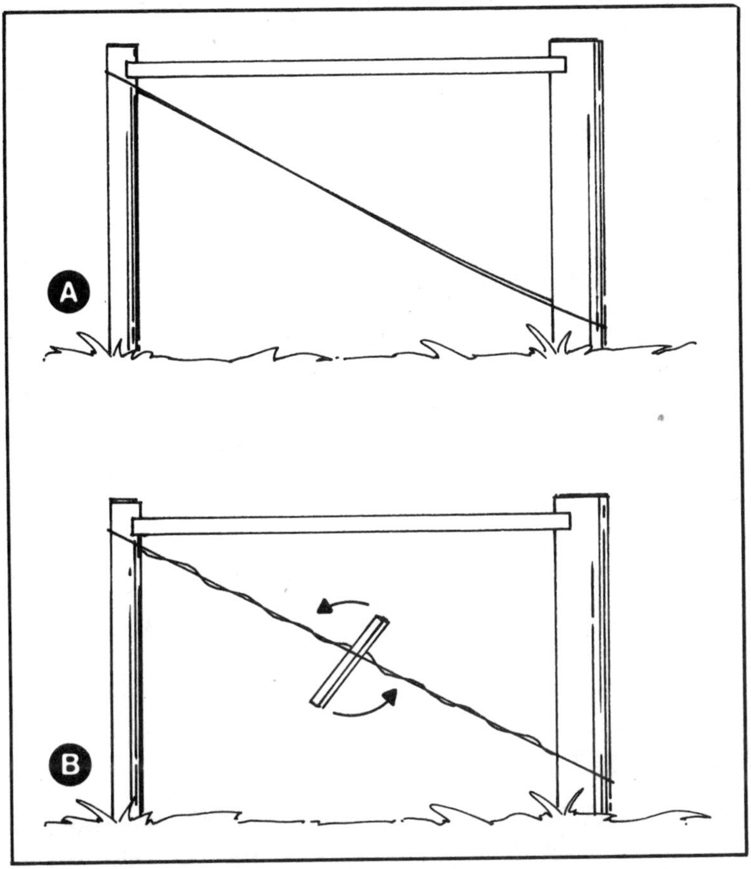

Fig. 5-11. How to tighten brace wire on corner post. (A) Loop wire around posts. (B) Insert a stick and twist until wire is taut.

INSTALLING WIRE ON THE FENCE

To stretch woven or welded wire, make up a wood clamp to hold the wire (Fig. 5-12). A double stretcher should be used for fence more than 32″ high, but many fence builders use just one.

If your goats will always be on one side of the fence, as a boundary fence, put the wire on the side with the goats so they will push it against the posts if they reach through it. For divider fences, put the wire on the side of the adult goats, or on the side where goats will usually be kept.

Unroll the wire so that it lies flat on the ground with the close mesh nearest the fence posts. Remove a vertical stay at the end and wrap each horizontal wire around the corner post at the proper

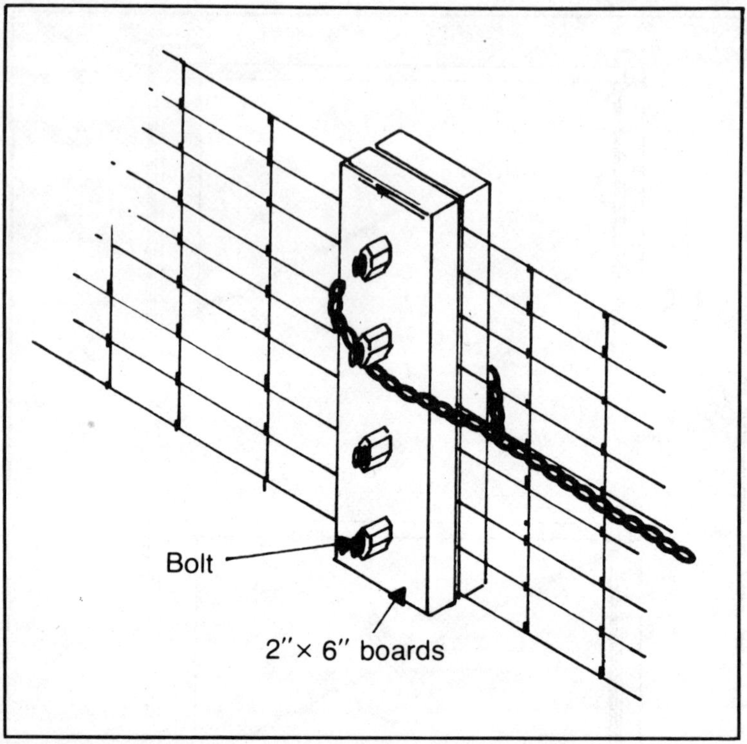

Fig. 5-12. How to make a clamp for stretching woven wire.

height. Tie it by wrapping the wire (Fig. 5-13). Staple each wire. Put the staples in at a diagonal so they won't be as likely to split the post (Fig. 5-14). Use 1″ staples for hardwood and 1-½″ ones for softwood.

Attach the clamp to the other end of the wire and connect the stretchers to it. Make sure your stretchers are securely fastened to the clamp and to a stationary object such as a tree or a dummy post. Stretch the wire taut enough so that the tension curves in the wire are about half straightened (Fig. 5-15). Go back along the wire and shake it to make sure it is free of any encumbrances. The wire should feel springy when you pull on it. Avoid exposing yourself unnecessarily to wire that is being stretched. Work on the side of the posts opposite from the wire as much as possible so that, in case of a sudden release or break of the wire, the posts would offer some protection. Wear heavy clothing and gloves, especially while handling barbed wire. Avoid sloppy fitting clothing that can catch on the wire and entangle you.

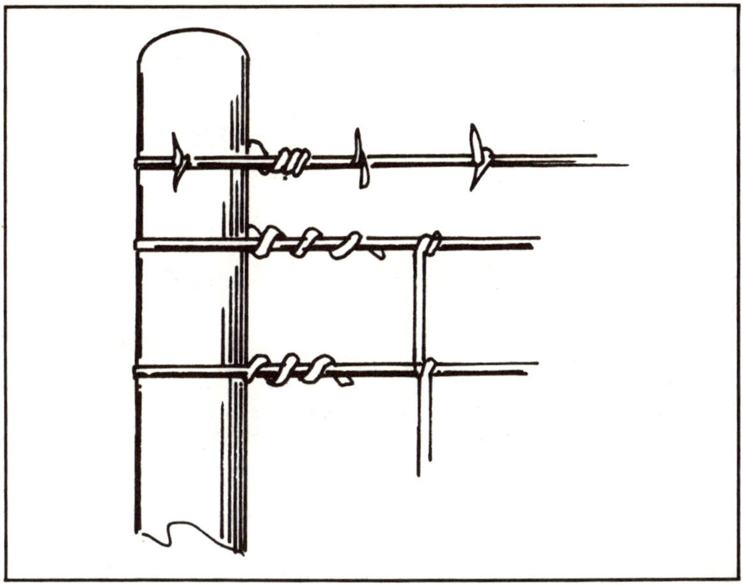

Fig. 5-13. Loop wires around end posts and twist them.

When you are satisfied that the wire is properly stretched, put at least two staples in each wire at the corner post. Move along the line of posts and staple the wire as you go, especially to posts that are on ridges or in depressions. After stapling, cut the wire to length and remove a vertical stay, using the horizontal wires to wrap the corner post. Never try to stretch a fence around a corner, unless the corner is an obtuse angle considerably greater than 90 degrees. You can splice a woven wire fence if necessary (Fig. 5-16).

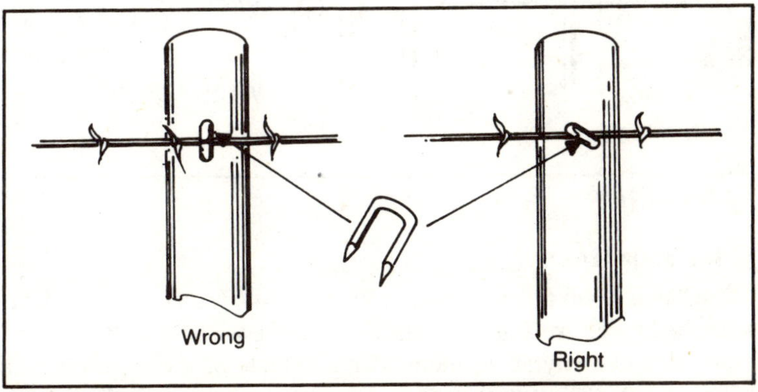

Fig. 5-14. Drive in staples at a diagonal so they won't split the post.

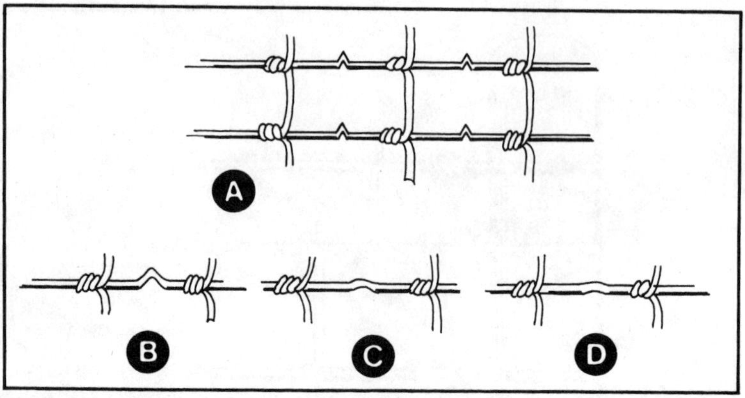

Fig. 5-15. How to know when a woven wire fence is properly stretched. (A) Woven wire before stretching. (B) Too loose. (C) Properly stretched. (D) Too tight.

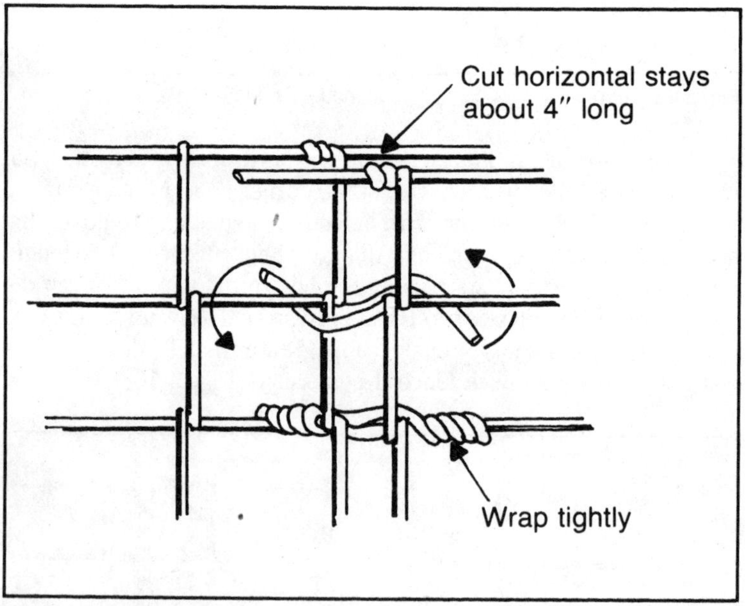

Fig. 5-16. How to splice woven wire.

FENCE MAINTENANCE

Check your fence regularly to prevent deterioration. Make sure the braces are solidly in place. Keep the wire tight. If a post or brace is broken by an accident with a vehicle or storm, replace it before the fence is weakened and the goats develop bad habits.

ELECTRIC FENCES

Some goat keepers prefer to use electric fences because they require a lower investment and are easier to move about. The typical electric fence for cattle has no more than a single wire, but goats soon learn to jump over the wire without touching it. If the wire is moved up, they go under it. Single wire electric fences are worthless.

The best electric wire goat fence is to use four or more wires, spaced every 6″, with the lowest wire set at about 1′ from the ground. This kind of fence works well, but it must be maintained properly. If the charger is turned off for any length of time, the goats will probably get out. When the charger is working, there is a clicking noise that may alert them to the danger of touching the fence. If you allow weeds to grow up into the wires, the fence's effectiveness will be decreased or completely lost by shorts.

Some owners use electric fencing as a protective device for woven wire fences. By spacing the electric fence a short distance inside the woven wire, the goats are kept off the permanent fence.

Electric fences are probably more useful as temporary fencing or protective fencing than for a permanent fence. They can be put up for much less money than a good permanent fence, but their reliability is questionable. If you don't mind tinkering with them and keeping the area near each one clean of weeds and tall grass, you might consider installing one.

Although people frequently say that goats will learn to respect an electric fence, that statement is not really true. One goat repeatedly went past an electric fence, knowing that he would be shocked. As he approached the wires, he would blink his eyes, showing that he knew what to expect. He went on through the fence. If you build an electric fence, you'd better hope you won't have such a goat.

OTHER KINDS OF FENCES

Farmers with a surplus of flat stones such as sandstone sometimes build rock walls. A loose wall can be built without a foundation. If you plan to lay up a wall with mortar, you must dig a trench below the frost line and pour a foundation. Such a fence is beautiful and permanent, but it requires a tremendous amount of labor to construct.

Living fences were used in earlier times. The Osage orange, also known as bois d'arc, goes by the name of "hedge" in some parts of the midwest because of the way it was used by pioneers. This thorny tree made a reasonably good fence for cattle and horses, in addition to providing posts for the farm fences. It has been largely abandoned for that use because the trees tap the moisture and nutrients in the soil near the fence, reducing crop production. The trees would have to be of good size before use, or the goats would probably denude them.

Multiflora rose is another living fence that can be used, but it has fallen from favor in many areas because of its habit of spreading into fields. It might work well for goats because they love the plant and would keep it out of pastures. The plants would have to be well established before exposing them to goats. A multiflora hedge is mainly useful to reinforce an old fence rather than to establish a new fence for goats. The combination of old wire, which is too rusty for goats, and the rose might work fairly well.

RESTRAINING GOATS

Sometimes you have a goat that is hard to retain. A goat that I called Mary was clever enough to work the latches on gates and doors. She would watch me close a door with an ordinary hook and eye latch. When I was gone, she would use her nose to push the hook back up out of the eye and release the door. She was equally inventive in the pasture. A livestock dealer came by one day and took Mary. I didn't want her to corrupt the rest of the goats.

Even if you install new wire, fences sometimes break. A goat that is determined to roam can exploit a minor gap in a woven wire fence, or a slight sag when a brace weakens, to escape. If you have an old rusty fence, the goat will make her own gate in due time. To keep her in and prevent the rest of the goats from wandering you must either repair the fence or, if it is beyond repair, build a new one. You can install a temporary restraint if that is inconvenient.

Tethers

One way to temporarily restrain a goat is to use a *tether.* This is practical only for goats that are rather calm. A nervous goat can become frightened and run so hard against the tether that she can be seriously injured or even killed.

Many people have used tethers for goats with no problems. Don't tether the animal near enough to a fence that she will be tempted to jump it. The tether can catch on the fence and the goat will be hanged. Don't tether a goat in the summer sun unless she can get to a shady spot. One doe was tethered in full sun on a summer day and began to run in crazy circles until the owner moved her. She

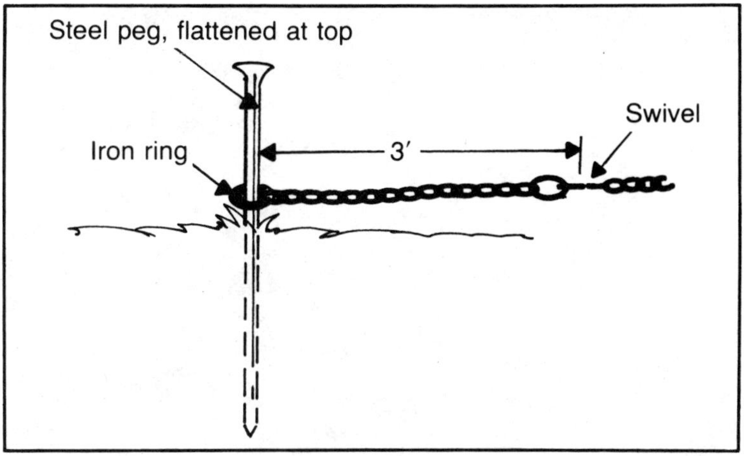

Fig. 5-17. How to tether a goat so that the chain doesn't wind around the stake.

was apparently close to heat stroke. If the tether stake is in the sun, the goat may wind the tether around the stake until she is caught fast against it. Put a swivel in the line about 3′ from the stake to reduce the chances of a tangled tether (Fig. 5-17).

You can use nylon rope for a line, but a more durable tether can be made from light chain. Go to a hardware store and ask for a twin-loop number 2/0 chain. This chain has a load limit of 250 pounds or more. For easy connection to the goat, put a heavy snap in the end of the chain that can be snapped into a ring on the goat's collar. Put a heavy ring at the other end of the chain. You may have to have a welding shop make a stake. The top of the stake should be enlarged or have a crossbar welded onto it, so the ring on the end of the chain won't slip over the top of the stake.

Nylon collars are reasonable in price and very flexible. New leather goat collars are expensive, but you can often find usable collars at flea markets or secondhand stores. Salvation Army stores usually have a collection of heavy belts, and some can be adapted to use for goats. These belts usually sell for about 25 cents. Get a D-ring from a hardware store that is just large enough for the belt to slip through it.

Tethers are far from the ideal way to restrain a goat, but with care you can use one. Some people use them after first obtaining a goat until there is time and money for a fence. You don't need a tether for every animal. A doe can be tethered, and her kids won't leave the area until they are quite large. This method of tethering by proxy can't be used if there are delicious young fruit trees just a few

Fig. 5-18. A tether should be used only as an emergency tool. When leading a goat, don't wrap the chain around your arm as the man is doing here. Have someone behind the goat, if possible, to encourage it to move.

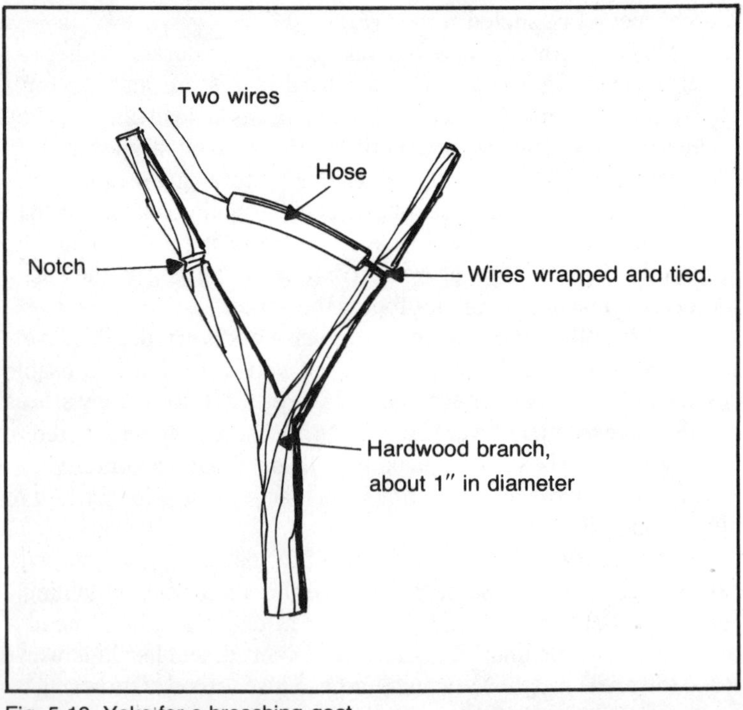

Fig. 5-19. Yoke for a breaching goat.

feet beyond the doe's reach. Try to use a tether as an emergency tool rather than a permanent way of keeping goats (Fig. 5-18).

Yokes

A *yoke* allows the goat much more freedom than a tether, and yet it is an effective way of keeping her inside a poor fence. There are precautions to observe. I know of one buck that was restrained by a heavy yoke which was held in place by bare wire. The wire gouged into the skin on the top of the buck's neck until it was raw. This happened during the summer months, so the inevitable occurred—screwworms in the wound. The owner noticed the buck lying in his house one August afternoon and went out to inspect. The

Fig. 5-20. A buck with a yoke. Use a lighter yoke for small does.

infestation of screwworms looked serious enough to kill the goat, but immediate treatment with a screwworm medicine, poured directly into the wound, ended the screwworms. The yoke was removed and discarded until the wound healed.

This problem of neck injury from yokes can be met in two ways: by using a light yoke and by seeing that no sharp edges touch the goat's neck. Rough bark or bare wire will rub the neck until the skin is broken. If you make a yoke with wire on it, cover the wire by passing it through a piece or rubber hose. Car heater hose is just about the right size and is tough enough to last. Measure the distance between the branches of the yoke carefully and cut the hose to fit exactly. There will be no exposed length of bare wire with no gaps.

To make the yoke, cut a branched section from a small tree that is about the right size to fit your goat's neck. The yoke should be about 18″ long. Notch each branch of the yoke just far enough above the crotch to accommodate the goat's neck. Select two pieces of smooth wire and wrap them around the notch on one branch. Twist the loose ends together at the outside of the yoke. Slip the piece of hose over the wire (Fig. 5-19). Place the yoke on the goat's neck. Wrap the wire and tie it at the notch on the yoke's other branch. Twist the loose ends of the wire at the outside of the yoke branch. Check that the yoke rests easily on the goat's neck and that no piece of wire touches the goat (Fig. 5-20).

Chapter 6

Feeds for Goats

Many people believe that it makes no difference what kind of feed you give a goat. Others admit that it does make a difference, but they say it's easy to make up a good ration for goats. Still other goat keepers consider the problem of providing a proper nutrition for goats one that requires intensive study. It can be either relatively simple or exceedingly complex, depending on whether you buy a mixed feed or mix your own. It also depends on just how far you want to pursue goat nutrition. While giving some guides for practical feeding here, I'll leave feed analysis to the biochemists who are employed by the feed companies. If you carefully inspect the list of ingredients in feeds manufactured by different companies, you will find variations indicating that even the experts don't agree on which mix is best for dairy animals.

Regardless of how you develop a ration, it's easy to tell if a goat is well nourished. An animal that has received good nutrition will not be fat, but it will be smooth and show a glossy coat (Fig. 6-1). A poorly fed animal will have a rough looking coat and may be poorly proportioned. It often has an enlarged middle. This condition is described as pot belly. An overnourished animal can be too fat, which makes it look more like a sheep than a goat (Fig. 6-2). A goat that is too fat lacks dairy character. Excessive fat can cause a serious health problem in pregnant goats, particularly just before or after kidding. Excessive fat may lead to *ketosis*—a pregnancy disease that can cause death.

Fig. 6-1. A doe that shows the effect of good nutrition. Note her glossy coat and excellent dairy character.

To determine the degree of fat in a goat, grasp the *flank*, the loose skin near the udder between the rear leg and the barrel. The thickness of this part will tell something about how much fat the goat is carrying. Also, check the size of the barrel. The barrel should be large. It should be deeper toward the rear of the goat. Feeding too little roughage to kids can restrict their capacity to eat and produce when they are adults (Fig. 6-3). This is why you should understand

Fig. 6-2. A victim of overfeeding. This Nubian yearling lacks dairy character, partly because she is too fat, even though she had been on a reducing diet 25 days when this picture was taken.

Fig. 6-3. This doe lacks dairy character, possibly because she was given a diet too low in roughage to adequately develop her paunch.

something of the anatomy and special nutritional requirements of goats.

GOATS AS RUMINANTS

Animals that chew their cuds are called *ruminants.* These mammals have a four-part digestive system, which many people describe as having four stomachs. The four organs do not all have the characteristics of a true stomach, so the statement is not anatomically correct.

The four separate parts are the *rumen*, which is the first and largest, the *reticulum,* the *omasum,* and the *abomasum.* To consider how the digestive system works, it is useful to divide the system into two parts. The first part consists of the rumen and reticulum, and the second part is the omasum and abomasum. The roughage that the animal eats goes to the first part of the system. It is regurgitated, chewed, and moved to the second half of the system.

The term "half" is somewhat misleading if you consider the size of the organs, because the rumen is the largest of the parts. It is sometimes called the paunch. When a goat browses or eats any kind of forage, she grabs a mouthful and quickly swallows it. The food goes directly to the rumen. When the rumen is full, the animal lies down in a comfortable spot and waits for the digestive system to work. After being processed into smaller pellets by the rumen and reticulum, the food is regurgitated. The pellets or cuds are chewed. When the doe swallows the chewed cud, it goes directly to the omasum for digestion, into the abomasum for further digestion, and on into the intestines.

Note that there is a certain efficiency here. A goat can go out on a hot day and browse for a short time. She can then retreat to the cool shade to chew what she has eaten. The rule for goats is eat in haste, digest at leisure.

You may wonder what controls the path of a goat's food, and why it goes to the rumen at the first swallowing and to the omasum after the cud is chewed. It depends on how the goat holds her neck when she swallows. When she stretches her neck, a valve in the esophagus opens to direct the food into the omasum. If she doesn't stretch her neck, the food goes to the rumen. Liquids are properly sent to the omasum rather than the rumen, so a goat stretches her neck to drink (Fig. 6-4). Kids must stretch to reach the dam's udder, sending the milk they get into the omasum. Although you can teach kids to drink from a bucket, they may drink without enough stretching if started too young and send milk into the rumen. This wrong way trip for liquid food can cause scours. When sugars must get into the bloodstream quickly, as when a doe has ketosis, the sugar can be dissolved in water and fed as a liquid. The doe will normally stretch to drink the liquid, sending the sugar solution directly to the omasum. This shortcut can save precious time and perhaps save the animal.

The rumen of an adult goat is large for the size of the animal and is often capable of holding 5 gallons of material. It is loaded with microorganisms that aid in food processing. The goat and her microorganisms are involved in a symbiotic relationship. The goat supplies food for the microorganisms, and they start the breakdown process that is necessary before the food can be digested. The coarse fibers that survive the first process are broken down by chewing when the goat regurgitates and chews her cud.

The rumen is an important organ in the digestive system. An athlete who plans to be a champion must exercise to develop his muscles even if he has inherited great ability. A young goat's rumen must be developed by consuming a diet that includes roughage. Wild goats get their forage from woody plants, but a dairy goat that has no access to living plants must have hay. This is why good quality hay is a prime requisite for a healthy goat.

Fig. 6-4. Note how this doe reaches to the opposite side of the bucket for water. This stretching sends the water to the omasum rather than to the rumen.

KINDS OF HAY

Hay can be divided into two main types—legume hay and nonlegume or grass hay. The most common legumes that are cut for hay are alfalfa and various kinds of clover. Legume hays are richer in protein than grass hays. Alfalfa has a protein content of about 15 percent, while the protein contents of grass hays are usually about 5 to 6 percent. The percentage of digestible protein in any hay is always lower than that of total protein.

Another advantage of legume hays is their calcium content. Calcium, which is necessary for good bone development, is supplied by legumes. Calcium and other minerals are supplied to the plant by soils that vary greatly in their available mineral content. Rainy wet climates produce soils of low total mineral salts content, because the salts are leached from the soil. An arid region will have soils of higher total salt content in the topsoil because of the constant evaporation from the soil. The water comes up from below the surface, evaporates at the surface, and leaves the salts in the soil. Soils in the southeastern part of the country are likely to be low in minerals, while those in the plains and western deserts are high.

It is debatable whether these differences in soils make a significant difference in the mineral content of legume hay, because alfalfa and other legumes don't do well in leached acid soils unless the soil is given additional lime. You can probably provide adequate calcium by feeding alfalfa even though you live in a lime-deficient region. Timothy hay grown on rich soil has a higher level of protein, carbohydrates, and ash than when grown on poor soil, and it is lower in crude fiber.

If you cannot buy good legume hay, but a grass hay such as Timothy is available, you can usually supplement the grass hay with alfalfa pellets. These pellets are made of compressed ground alfalfa. They vary in food content according to how the water is removed—in some brands by sun-drying the alfalfa. In the production of other pellets the alfalfa is partially dehydrated by artificial means. The second kind of drying is more effective and produces a pellet with more concentrated food value. These pellets can have up to 17 percent or more crude protein.

HAY QUALITY

If you want your goat to eat the hay you buy, the quality of the hay is even more important than the kind. Goats refuse dusty or moldy hay unless they are starved to it. Some people run the hay through a baler before it has cured sufficiently, which produces a bale that contains too much moisture. After the hay is made into a bale, it cannot dry adequately. Bacteria and molds can grow. If you put many wet bales of hay into your barn, directly from the baler, they will become hot. Hot hay can burn a barn. Hay that is baled with enough moisture left in it to cause heating can never cure properly and will be of low quality.

Sometimes the weather during hay harvesting makes for poor hay. If hay is cut and rain falls on it while it is curing, the water can leach the nutrients from the leaves. Even if the hay is left in the field until it is dry, it will be of poor quality. If the hay is baled before it is thoroughly dry, and this often happens because people fear that the hay will get wet again if left out too long, molds and bacterial growth will destroy the hay's carbohydrate content. This hay is of limited nutritional value, and it is so unpalatable that a goat will eat very little. If the goats have no other hay, they eat little and what they eat is of low nutritional quality. The goat can starve to death even though plenty of hay is available. This is particularly true in cold climates where insufficient grain or other concentrate is fed.

The problem most goat owners face is that they have little firsthand information about how a batch of hay was processed. The ideal way to get hay is to pick it up in the field behind the baler. You must make contact with the grower before the hay is harvested and monitor the weather during harvest. If you must buy hay without such information, check the bales to see if the hay is bright or dull. If the leaves in the hay have a dull surface, buy one bale and break it open so you can inspect the hay in the center. Slap it to see if dust or mold is present. Check the color. Cured hay should have a greenish cast rather than a dirty straw color. Reject any hay that fails these tests unless you can find no other.

Some hay gets too much curing. Alfalfa hay that has been left too long in the field before baling will shatter when the baler picks it up. It loses most of its leaves. When you buy such hay, you get mostly dry stems with low nutritional quality. The hay won't hurt your goats if you supplement it. Most people who put up hay try to avoid overcuring, but many interruptions can occur with experienced hay harvesters. Balers can break down, people get sick or tired, or a birthday or anniversary can intervene. Inspect the bales before you buy.

Another factor that influences hay quality is the time of cutting. Hay must be cut at its prime to be high in nutritional quality. If the hay is left growing too long until the flower buds appear, it will be stemmy. Such hay is higher in fiber and lower in nutrients than hay that is cut earlier. Again, late cutting occurs with even the best farmers when other work interferes or problems arise. You can identify such hay by its appearance. When you break open a bale, the hay looks much coarser than top quality hay.

If your goats are not on pasture, the vitamin content of their milk will be affected by the kind of hay they get. Milk normally has a higher vitamin A content in summer when lush pastures are available than in the winter when there is no pasture. Goats that have access to pasture produce milk with a higher vitamin A level than those that never see a pasture. Green pastures are a good source of *carotene*, the raw material for vitamin A. Hay made under ideal conditions will have a higher carotene content than hay that is carelessly handled. Well cured, leafy alfalfa can have up to 10 times the carotene content of hay that is too stemmy or of otherwise poor quality. The carotene level is affected by the time of harvest and the length of time required for curing. Hay or pellets produced by artificial dehydrating will be higher in carotene than hay that is only sun-cured. This is another reason for feeding artificially dehydrated alfalfa pellets. These pellets are a good supplement for any hay feeding program that includes poor quality hay.

Find out whether any poisonous sprays were used by the grower on the hay field or on adjoining fields just before harvest. Most hay producers use reasonable care to avoid such contamination, but not all. Hay can become contaminated, either in the field or during transport, through no fault of the producer. This problem is probably less serious now than in earlier years when it was largely ignored through ignorance. Knowing your hay grower is a great advantage.

See if the hay contains a significant amount of weeds. Some weeds affect the taste of milk, while others are just unpalatable to goats. If you are buying alfalfa hay, make sure you get alfalfa and not a cheap mixture. Clover hay often contains a mixture of grasses and clover. This hay is satisfactory if properly processed, but know what you are getting. Mixed hay should be available at a reduced price to the going rate for good alfalfa. Sellers have a habit of calling any hay that contains a bit of alfalfa "alfalfa hay."

Some goat keepers believe that alfalfa is too rich for goats. They feel that a high protein hay burns out animals. There is little experimental evidence, if any, that this is true. The only problem is to make sure that goats get enough roughage.

HOW MUCH HAY?

The quantity of hay that one goat will consume depends on the availability of pasture. Many goat keepers prefer to feed hay as the only forage year around, believing that they get higher quality milk by eliminating the chance of strong tasting wild plants in the diet. This may be a consideration for you, but it will raise the cost of the milk you produce.

As a conservative estimate, plan to supply each doe with a ton of hay per year. This is a maximum figure; you can probably get by with much less. It also assumes that no pasture is available. If you have a ½ or even ¼ acre of pasture per doe, you can reduce the amount of hay required to ½ ton or less, depending on the length of grazing season in your climate.

PASTURES

When I told the old woman who sold me some goats that I had plenty of pasture for them, she looked at me skeptically and said, "Well, goats don't like sweet grass." I asked what kind of grass she meant. "The kind that cows like," she replied.

I learned later what she meant. When I released the goats into my lush pasture, they took a few bites of grass, then made a direct path to some multiflora rose that was growing in the fence line. They then wandered to the stream that runs through my pasture, nibbling at tree leaves and brush along the banks. In a short time all trees were shaped like huge toadstools, with the undersurfaces trimmed flat at the heights the goats could reach by standing on their rear legs. Goats prefer browse.

Pasture Management

Goat keepers frequently warn novices not to depend on short grass pastures, because goats don't do well on such forage. These pastures often cause more trouble if allowed to get too lush and if located on low ground in a damp climate. If this is the only pasture available and it is small, the pasture should be divided so that the goats can be rotated from one part to another for short periods. This practice reduces the chances of parasite buildup on the ground.

Another way to use grass pasture, especially one that is in a damp area, is to follow the European practice of bringing the pasture to the goat. By cutting the grass and delivering it to the pen or barn, you can control the goat's diet and eliminate the strong tasting weeds or grasses that would flavor milk (Fig. 6-5). How much green forage should you feed? A goat should have about 3 pounds of dry hay per day. Green forage has a high water content, so much more of it will be required. The water content may vary according to climatic conditions and the kind of forage, but you will probably need at least about 10 pounds of forage per day. This is one way to force your goat to eat "sweet grass" that she would refuse if she could roam and choose her food.

In addition to controlling the diet, cutting green forage plants also conserves pasture. If the pasture area is very small, goats will soon refuse to eat the plants that they have walked over and contaminated with manure or urine. Goats will rarely eat plants that grow in small pens. The pens are often bare of green plants, but only because the goats have trampled the grass with their hooves. Another way to conserve pasture in a small backyard lot is to tether the goat and move her each day.

Kinds of Pastures

Although legumes make good pastures for goats, there is the danger of bloat. Alfalfa and ladino clover pastures are among the most dangerous, especially if no grass is mixed with the legume. If

Fig. 6-5. One way to efficiently utilize pasture is to cut and move it to the goats. You can put up small amounts of hay by hand, but loose hay requires a storage area many times larger than that required by baled hay.

bromegrass, Timothy, or orchard grass can be mixed with the legume, the legume will supply the extra nitrogen needed by the grass for good growth. These mixtures make the pasture less likely to cause bloat, but they are no guarantee that bloat won't occur. One way to prevent bloat is to expose goats gradually to a lush pasture. Begin by bringing a small amount of the clover or alfalfa to the goat each day, increasing it gradually. Before turning goats into the pasture, let them eat their fill of dry hay. Alfalfa is especially likely to cause bloat during seasons of rapid growth. You should not let goats into the pasture after cutting the hay, unless they are introduced to it for a short time each day. The goats need to be accustomed to the rich forage, so they are less likely to gorge on it.

Tall *fescue* is one of the most durable pastures and can survive drought or cold winters. Fescue spreads, resists the growth of weeds, and holds soil against erosion. It also grows thick and tall and produces a large yield. The nutritional quality of fescue is much lower than that of legumes, and it isn't very palatable, especially during summer. Because fescue will grow under adverse conditions, it will supply much needed pasture when other plants will be almost dormant.

Some goat keepers praise *comfrey* as a green forage plant because of its high protein content. In some climates it can have a nitrogen content that is too high for safety. Apparently this condition can be caused by either a cool rainy season or a hot dry one.

Pastures for Young Goats

If you separate the kids from the adults, there is no reason to restrict the kids' diet. If they can browse on woody plants, their fiber content will aid in rumen development. The minerals that woody plants bring up from the subsoil are different from those available in shallow-rooted plants. You can use young kids to control weeds and brush anywhere on your property. Some goat owners have used them to control weeds outside property lines on streets or country roads. If you do this, make sure that the area is free of poisonous sprays. Telephone personnel sometimes spray herbicides under rural telephone lines to kill the vegetation. Move your goats while this spraying is done, and keep them away from sprayed areas until sufficient time has passed to reduce the danger. Goats are susceptible to herbicides, so you should avoid a sprayed area until all signs of the spraying are gone. The most dangerous time is immediately after spraying, before the effects are visible on the plants.

Another danger in roadside plants occurs where traffic is heavy enough for the plants to be polluted with lead from car exhausts. Lead poisoning is less likely from this source now than when all gasoline contained lead.

CONCENTRATES FOR GOATS

There are at least two easy ways to choose a suitable concentrate for dairy goats. Feed either a special goat feed like Purina Goat Chow or any good dairy cow ration. Check the label to see if urea is an ingredient. If so, choose another feed. The use of a special goat feed assumes that it should not only be available in your area, but that it is used by enough goat keepers so that the feed store will have a constant turnover of fresh feed. Feed that has been stored too long can become moldy or infested with various insects. Such feed is neither palatable nor healthful for goats.

You can let the goat decide on the best ration to some extent. A few goats will almost refuse to eat any pelleted feed. It may be possible to starve them to it, but that could create health problems before it worked. Finicky eating can in itself imply a health problem, but it is more likely a matter of long habit. A goat that has been fed a diet of crimped corn and oats may reject a pelleted feed that includes molasses. If you want to change a goat's ration, do it slowly.

Regardless of the concentrate you decide to use, make sure it is a coarse mixture rather than fine. Goats do not like grain or other concentrates that have been processed to a dustlike particle size. They like whole grains and grains that have been crimped or cracked but not ground.

Commercial or Home-Mixed Feeds?

Although you can mix a ration from grains and supplements, it probably saves you little if any over a ready-mixed dairy feed. The only way you can make substantial savings is by growing your own feeds. This implies that you have the space and tools to do it. One problem with home-mixed feeds is to provide sufficient levels of trace elements such as copper, cobalt, manganese, and zinc. These elements are supplied by ready-mixed feeds and are a good reason for using such feeds, unless you live in an area where the soils are rich in trace elements. Even where the trace elements are available in the soil, there is no guarantee that the plant will contain a satisfactory level of the element. The availability of such elements varies with the chemical makeup of the soil. Lime, which is used to raise the pH of soils so that alfalfa can be grown in them, depresses the intake of cobalt. Get a dairy feed to insure the availability of trace minerals. You can use a mineral salt supplement to supply trace minerals, preferably a mineral salt block.

Grains as Concentrates

Some goat owners feed nothing more than oats or oats plus corn to their dairy goats. Others claim that an oats diet will adversely flavor the milk. Where such feeds are used, keep the corn content to about one-third or less of the total ration, because corn usually contains no more than about 8 or 9 percent crude protein. More corn than this will reduce the protein percentage of the mixture too much for a milking doe. If sufficient protein is your only concern, you can feed oats and alfalfa pellets and probably fill the need. If you are feeding a grass hay, you will need a higher protein concentrate than straight oats. Oats contain about 12 to 13 percent total protein, but only 8 to 9 percent digestible protein. One way to feed oats and yet supply enough protein is to add a small amount of high protein commercial calf supplement such as Carnation Calf Manna to the ration. Don't forget to provide a mineral supplement.

Concentrates for Kids

As soon as a kid will eat dry food, you can supplement the milk ration with a calf starter (Table 6-1). Purina states that a kid will eat about 100 pounds of their Calf Startena while growing from one week to four months old. This kind of feed is often available in either a coarse or fine mixture, but goats like the coarse mixture better.

Don't overfeed young goats. Too much grain will make the goat fat and, by depressing the appetite for roughage, will restrict rumen development. The yearling doe shown in Fig. 6-2 was raised on a ration of about 3 pounds of concentrate per day. When Fig. 6-2 was taken, she had been on a reducing diet of 1 pound per day for 25 days. Note the heaviness in the shoulder region. This may be partly caused by heredity, but much of it was produced by overfeeding. It is permissible for Nubians to show a bit more fat than other breeds, but this yearling shows far too much. Look again at Fig. 6-1, also a Nubian, to see what a good dairy character looks like. The only defect in this doe is the size of the udder, but she had been milked just shortly before this photograph was taken. You can assume that she was fed a well-balanced ration of concentrates and roughage during her developing months. She has the angularity of body associated with good dairy character. The fat yearling looks more like a beef animal. She was likely fed a concentrate ration that was not only too abundant but poorly balanced.

How Much Concentrate for Goats?

Refer to Table 6-1 for a guide to feeding kids and young goats. Table 6-1 also gives a rough guide for feeding milking does, with a note at the right about challenge feeding. Some goat owners use this system to arrive at the proper amount of feed for each milking doe. To use challenge feeding, they increase the amount of concentrate each day while carefully measuring the milk yield. If the milk yield increases, they add a small amount of concentrate to the feed ration. When the milk yield levels off, with no more increase even though the ration is raised, they maintain the feed quantity until the milk production declines. These changes must be gradual.

The advantages of challenge feeding are that you tailor the amount of feed to the individual doe, and you get maximum production. Another way to fit the amount of concentrate to the doe is to weigh the milk and feed a certain ratio of feed to milk weight. Purina suggests that you feed 1 pound of their Goat Chow to every 3 pounds of milk.

Table 6-1. Feed Schedule for Goats.

Goat	Pasture	Kind of Feed	Amount	Comments
Kids to 30 days	Not necessary.	Milk. Offer a coarse calf starter. (Purina Calf Startena, for example).	½ pint at first, then 1 to 1-½ pints.	Make sure new kids get colostrum.
Kids 1 to 4 mos.	Browse, Free choice.	Milk. Calf starter (as above). Or grain.	1 to 1-½ pints (milk). Free choice other feeds.	Kids prefer a coarse feed to pellets.
Young goats 4 mos. and over	Browse, Free choice.	Dairy ration (Purina Goat Chow, for example). Or oats.	1 lb.	Supply enough roughage to develop rumen.

Dry does	Browse, Free choice.	After 100 days, give dairy ration.	1 lb.	Feed should contain molasses
Milking does	Controlled, no wild onion, etc.	Dairy ration.	1 lb./3 lbs. milk.	Can use challenge feeding to set proper level.
Bucks	Browse, Free choice.	Dairy ration. Or oats.	2 lbs.	
All goats				Free access to legume hay and mineral salt.

People who feed straight grain are usually less specific in the amount of concentrate they give a milking doe. One owner told me that he feeds about 1 quart of oats per average doe. That is not very specific.

Precautions in Feeding Concentrates

You can do worse than follow the method of 1 quart a day. How? You can feed so carelessly that you don't know how much feed each animal is getting. The amount of concentrate should be fairly constant. Quick changes are not for goats unless you are prepared for serious trouble. Any sudden change in the amount or kind of feed can upset a goat's digestion and production.

Some goat keepers mix their own feeds, and this can be tricky if you get too fancy. Soybean meal is often added to grains to raise the protein level. The precaution with soybean meal is to never mix it with a feed that contains urea. You should avoid any dairy feed mix that contains urea anyway. It isn't recommended for goats. If you do mix your own feeds, you can adjust the optimum protein level by adding a high protein concentrate such as special calf feeds. The Carnation Company produces a feed called Calf Manna that supplies 25 percent protein. A measuring cup included in each bag of Calf Manna holds enough feed for one day for three goats. You can use the cup as a guide to maximum feeding of this concentrate. Begin with a tablespoonful and increase the ration by challenge feeding until milk production peaks. Don't overfeed a protein supplement.

Choose a feeding concentrate with molasses for dry does. This helps prevent ketosis. If you are feeding oats, give the doe molasses in warm water just after kidding.

If you keep insecticides or pesticides anywhere on the premises, make sure they aren't handled near the goat feed. Any contamination by such agents is dangerous for the goat and for anyone who drinks the milk. Some horticultural chemicals can be stored in the animal's fat and released through the milk for months after the time of exposure.

There are many ways that toxic substances can get into the food system. One way is by using discarded chemical containers for watering tanks. Plastic containers are likely to hold traces of contaminants even after careful washing. Find out what was in these containers before you expose your goats to them. You can get used plastic buckets that are perfectly safe and low in cost from restaurants or doughnut shops. These establishments buy their pie or doughnut filling mixes in 4-¼-gallon plastic containers and will

Fig. 6-6. Don't serve water in a bucket with a bail as this man is doing. When goats get their heads caught in the bail, they become excited and can be injured.

usually sell the empty buckets for about $1 or so. These buckets have tight-fitting lids that make them useful for storing small amounts of feed supplements. They are much more durable than the typical cheap plastic bucket that you would pay more for at a discount store. Remove the bail from a bucket before using it as a watering tank. Goats that get their heads caught in a bail become excited and can be injured (Fig. 6-6).

WATER QUALITY

A man once told me that goats would not drink clean water. He said they preferred water contaminated slightly by dirt or other material. That is an unfortunate misconception. There is no substitute for fresh clean water, especially for your milking does. Water is the top ingredient of milk, and the doe must have an abundant supply of clean water to produce at her peak level. Goats that are not milking can survive on little or no water for considerable periods, but it's not recommended. Make sure all your goats have access to clean water.

The temperature of the water becomes important during cold weather. At subfreezing temperatures most goats will drink very little water, unless the water is heated to about 60 degrees Fahrenheit or higher. One easy way to provide goats with warm enough water is to keep the water in your house until it reaches room temperature. By using a couple of large buckets, you can supply them with water that is much warmer than the air at least twice a day. Make the water available to them before you give them grain. The importance of warm water is hard to overemphasize. Goats will not only drink more warm water, which is desirable for their production, but they require less feed for equal production when given warm water. If the goat drinks water that is at the freezing temperature, her body must supply heat to warm the water. That heat comes from metabolism of food.

If you give your goat cold water, you must pay more for the milk she produces by feeding extra calories. A calorie of heat is required to raise 1 gram of water 1 degree Celsius. If the air temperature is very low and the doe drinks even a few ounces of water per day, at 28 grams to the ounce, the number of wasted feed calories can be significant. The story is the same even with dry does or bucks. Animals are under strain during cold weather anyway, so don't aggravate their problem by giving them icy cold water. If you do that, their usual response is to drink little or no water, and that will affect production. Even with warm water most goats will drink far less water in winter than in the summer, but make sure that water is available for those who want it.

Goats will not drink if droppings or other obvious contaminants get into their water supply. This is one reason why a plastic bucket makes more sense for the average small goat herd than a fixed watering tank that is hard to clean. Some goat keepers use an automatic water device that supplies fresh water when the goat licks the fitting at the end of a pipe. These devices supply clean water on demand, but they are a problem during freezing weather. Another disadvantage is they drip water onto the ground and make a wet, muddy area. One solution is to build a box around the device, much like a grain feeder, so that goats can't walk or lie down in the mud.

Chapter 7

Goat Hygiene

Goats practice good hygiene by instinct. When a goat is kept under partial restraint in a pen or shed, you must provide some necessary services under those conditions. Wild goats may need their horns for protection from predators, but horns are a nuisance in close association with other penned goats. Goats that clamber over rocky ledges and other rough terrain keep their hooves in good condition, but goats on land with rich top soil grow long "toenails" that inhibit walking and can invite disease. Goats in the wild grow shaggy hair for warmth, but long hair in a dairy goat may lead to unclean milk.

In addition to providing grooming care for your goats, you must eliminate surplus male kids. There is a limit to how many functioning bucks the world needs, so the logical way to use extra males is to castrate them as kids and grow them for meat. Some owners find this hard to do, but it makes sense unless you are a vegetarian.

Even if you have never performed any cosmetic or surgical modifications on animals, these tasks are easy to learn. You should not leave them for a veterinarian; that would be inconvenient and expensive.

HOOF TRIMMING

If you don't set up a regular schedule for hoof trimming, the hooves will become overgrown, then collect manure and other debris that cause foot rot. The timing of your schedule will depend partly on the terrain where you live (Table 7-1). Rocky pastures will

Table 7-1. Suggested Herd Management Chart.

Name	Hoof Trim	Body Trim	Wormed	Next Worming
Maisie	1/10/83	2/15/83	6/17/83 TBZ	12/83 tramisol
Cleo	1/10/83	4/5/83	6/17/83 TBZ	12/83 tramisol

wear hooves to some extent, but trimming will be required even in an area with rocky soil. The more frequent the trimming, the less cutting you will have to do and the better for the goat. If the trimming is delayed too long, there is more danger of causing bleeding. It becomes necessary to remove some of the pad along with the outer hoof shell to make the hoof bottom level.

Hoof trimmers often use a sharp pocketknife or linoleum knife, but pruning shears are safer for both the trimmer and the goat. If you buy a pair, get the curved-blade type instead of anvil shears.

The goal in hoof trimming is to make the bottom of the hoof level and parallel with the top. Some trimmers claim that they can do this better with a sharp knife than with pruning shears. This may be true in theory, but when you start whittling on a moving hoof, who knows what will happen? To reduce the chances of injury, use some kind of restraint for the goat such as a collar and chain, a stanchion, or even people holding her (Fig. 7-1). Trying to trim hooves with no restraint can be dangerous, especially for the goat. If she tries to jump away while you are holding her foot, she may break a leg. Plan to do the trimming at feeding time if possible, so you can use a pan of feed to pacify her. Try to work against the wall of a building so she can't move sideways. Gentle words and slow movements will help persuade her that you mean no harm. The surgery is generally painless if done frequently and carefully.

Pick up a front hoof and bend the leg at the knee, so you can bring the foot up between your legs the way a blacksmith shoes a horse. Hold the hoof with one hand. Brace the hoof against your leg and do the cutting (Fig. 7-2). Start cutting at the heel and cut toward the toe. If you are using pruning shears, slip the narrow blade along the hoof with the end of the blade toward the toe (Fig.7-3). Do one toe completely (Fig.7-3), then the other one. Try to get the two toes trimmed equally so that when the goat walks, her weight will be distributed equally on the toes (Fig.7-4). The trimmed hoof will look significantly different from the untrimmed one if the hooves have been neglected (Fig. 7-5).

Fig. 7-1. When trimming hooves, always divert the goat's attention with feed. Use some kind of restraint such as a stanchion or someone holding her.

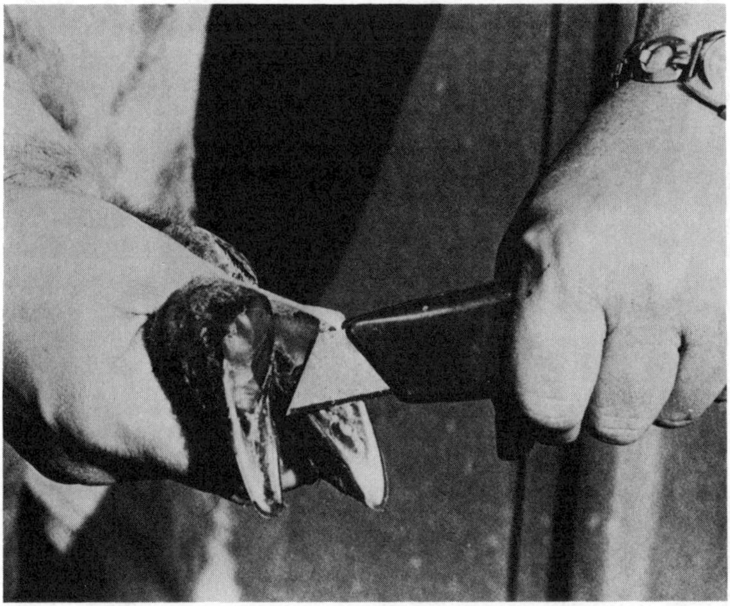

Fig. 7-2. The direction of the cutting stroke should be from heel to toe.

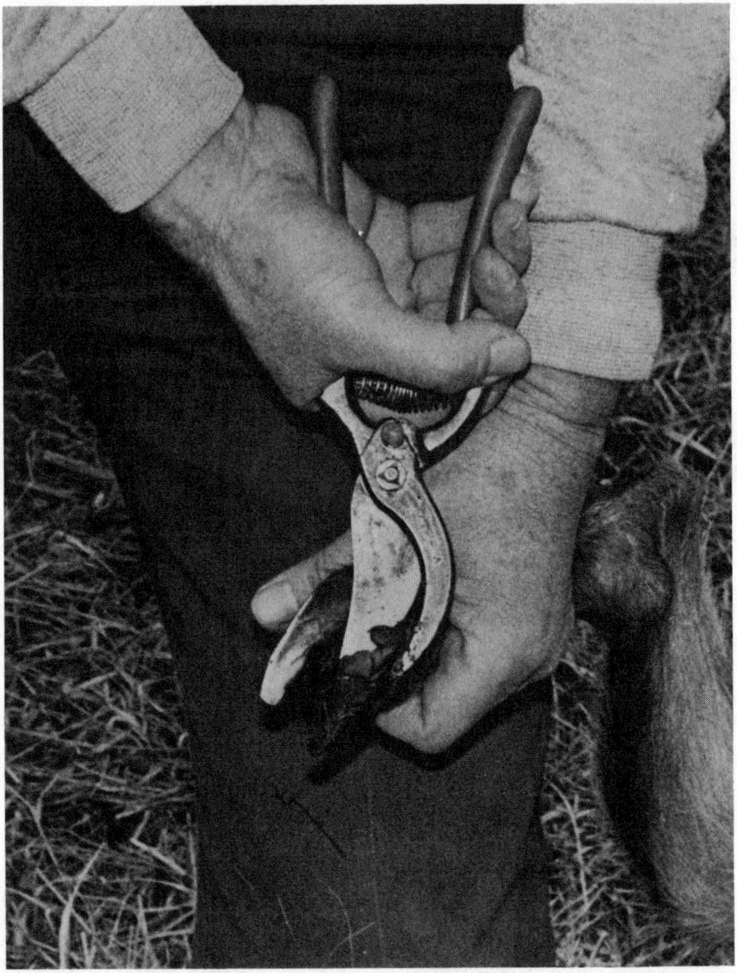

Fig. 7-3. Pruning shears make for safe work in hoof trimming. Note how the untrimmed toe has overgrown and been turned under.

When you cut off any overhanging hoof shell, you may find foreign material in the hoof. Use the point of the pruning shears or knife blade to dig it out. An offensive odor will likely be present because of bacterial growth in the filth that accumulates in overgrown hooves. Use a hoof dip after trimming. You can mix 1 pound of bluestone crystals (copper sulfate) to 6 gallons of water. In metric measure use 20 grams of bluestone to 1 liter of water.

After trimming the front hooves, follow the same procedure for the rear ones. Rear hooves are especially important. While a goat

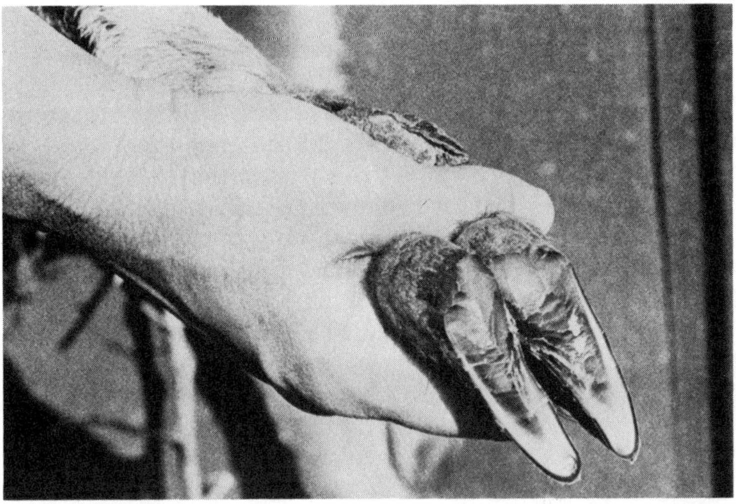

Fig. 7-4. The appearance of a hoof after trimming. Some trimmers prefer to make the toes shorter than those shown here, but stop when the cut area becomes pink.

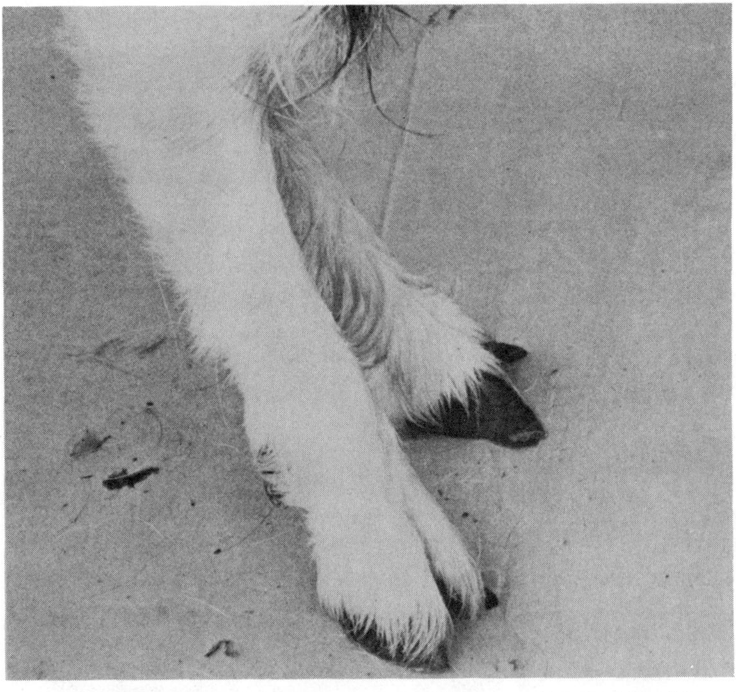

Fig. 7-5. Note the difference in stance between the trimmed hoof on this side and the untrimmed hoof.

can graze on its knees when there is a problem with the front feet, it is rendered immobile by a serious problem with the rear hooves.

To recap hoof trimming, check that you have removed all the folded edges of the hoof shell, which forms cavities for foreign material to collect. Are the hooves clean? Has all excess material been removed so that a white line shows around the edge of the hoof? Have you leveled the hoof by removing a thin slice of sole? This last step may be unnecessary if you trim often. See how the goat stands on the hoof. Does she stand up straight without leaning forward? Does the hoof turn up in front? If the goat leans forward when standing more material must be removed from the heel. A hoof that turns up in front needs more trimming at the toe. You can do minor shaping with a wood rasp or similar tool. If you cut deep enough to bring blood, stop and do the next hoof. A light pink color is okay, but try to avoid deep cuts that can cause bleeding.

HAIR TRIMMING

Hair trimming keeps the animal cooler in summer, improves her appearance, and aids in producing clean milk. The udder area should be trimmed close to reduce the possibility that goat hair will get into the milk. Hair in the milk is one cause of off-tasting milk. If the hair around the udder is long, it will hold manure and other forms of contamination even after scrubbing with soap and water. Animals to be shown at a fair must always be neatly trimmed. The typical time for trimming is spring—about kidding time. If a doe kids during winter, trim the udder, but leave most body hair for warmth.

The ideal trimming tool is a set of electric shears designed for shearing animals. Some shears are rather expensive, so you might check your local flea markets for a used set. You can use shears designed to cut human hair if you find a used set at a low price. The shears will be slower to use, but they will do the job if you don't rush. Animal shears work faster and will probably last longer. You can buy a hand-operated animal clipper at a significantly lower price than the electric models.

Trimming will require some practice. The goal is to make the animal as smooth as possible (Fig. 7-6). A goat that has been trimmed well appears much smoother than one that has a rough, untrimmed coat (Fig. 7-7). Use care in trimming so that you don't gouge the skin. If the job is done carefully, your goat won't mind it. If you dig the points of the clippers into her hide, neither of you will enjoy the task.

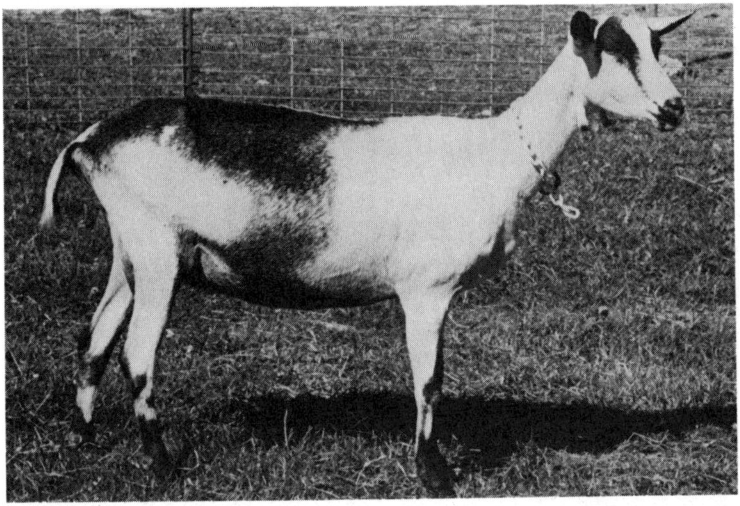

Fig. 7-6. This clipped goat has a smooth, clean appearance.

DISBUDDING

As mentioned earlier, horns are a nuisance and a danger for dairy goats (Fig. 7-8). Some buck owners will not accept does for breeding if they have horns, fearing that their bucks might be injured. Dehorn a goat when it is only a few days old. Dehorning adult goats is a difficult job—one that veterinarians don't like to do and often won't recommend. The time of healing for an adult goat is such that infection can enter the wounds, and you can have a dead goat. The horns are hollow. Cutting them off after they have grown out will leave an opening for organisms to enter.

There are two widely used methods of disbudding—*chemical* and *heat*. The chemical method is done by painting the horn bud with a caustic solution that destroys the horn tissue. The other method, burning, is done with a hot iron. Some owners use an iron that must be heated to the right temperature in a fire or blowtorch, but modern irons are heated by an electric current. An electric iron resembles a soldering iron except that the end of the iron is blunt rather than pointed.

The hot iron treatment is quicker to apply, but it can be dangerous if the iron is held on the horn bud too long. The iron costs more to buy than a tube of dehorning paste, but it can be used indefinitely without more expense. The dehorning paste has its dangers; any paste in the goat's eye can damage vision. You must

Fig. 7-7. Shaggy hair not only looks bad, but it makes sanitary milking more difficult.

make sure other goats don't lick the paste from the goat's head. Caustic materials don't improve rumen performance. Disbud as soon as possible after you can feel the horn buttons.

Caustic Paste Method

Dehorning paste is usually sold in plastic bottles or in applicators. The active ingredient is *caustic soda, sodium hydroxide,* or *caustic potash* (potassium hydroxide). These agents can cause severe burns. If you get any paste on your skin, wash quickly using warm water. If possible, submerge the affected part in water, then rinse the area with vinegar. If any goats or humans swallow the paste, give them a generous dose of vinegar or citrus juice— lemon, grapefruit, or orange. Follow with a dose of olive oil. If the paste gets into an eye, wash the eye with a 5 percent solution of boric acid. Quick treatment is essential, and professional medical help should be obtained.

Clip the hair from around the emerging buttons. Smear a layer of *petrolatum* on the skin surrounding the buttons. Make sure the buttons are free of material that might interfere with the disbudding paste. Squeeze a thin film of paste onto each horn button (Fig. 7-9). Use care to prevent getting the paste on other parts of the kid or on you. Cover the treated horn buttons with some short pieces of masking tape; then wrap the kid's head with a strip of tape to hold the short pieces in place (Fig. 7-10). The masking tape prevents the kid from rubbing the paste away from the buttons, and it also protects other kids that might try to lick the treated area. Leave the masking tape in place for several hours. Protect the kid from wet weather for at least 12 hours after treatment. When you remove the tape, don't be surprised if the area around the buttons looks sore. Neutralize any excess caustic by swabbing the area with vinegar.

Fig. 7-8. Horns are a nuisance and potentially dangerous.

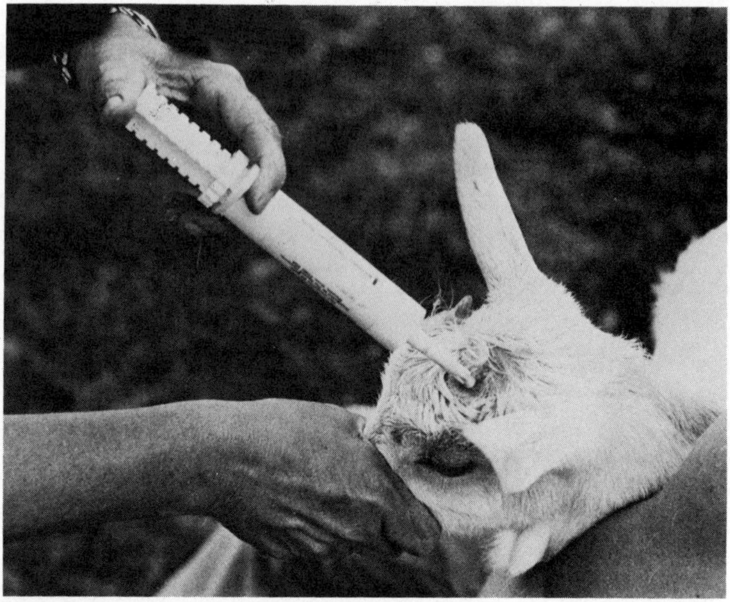

Fig. 7-9. When dehorning with caustic paste, squeeze a thin film of dehorning paste onto each horn button.

Fig. 7-10. To keep the kid from rubbing the paste off or other goats from licking it, cover it with masking tape.

Hot Iron Method

A 1″ iron is best, but a ⅞″ iron can be used with many kids. Do the job within a few days after the kid is born. If you wait too long, the job won't be successful.

Beginners often become impatient and try to use the iron before it is hot enough. If the iron is too cool, it will do little except cause pain for the kid. The iron should be cherry red at least 2″ from the tip. Wait until it reaches that temperature. Some inexpensive disbudding irons don't get hot enough for painless disbudding. Obtain a good electric iron or use one that can be heated by a blowtorch or hot coals.

Disbudding will require only a few seconds if the iron is right. Hold the kid steady by placing its forelegs between your legs. Press the iron down on the horn button (Fig.7-11). Rotate the iron slightly

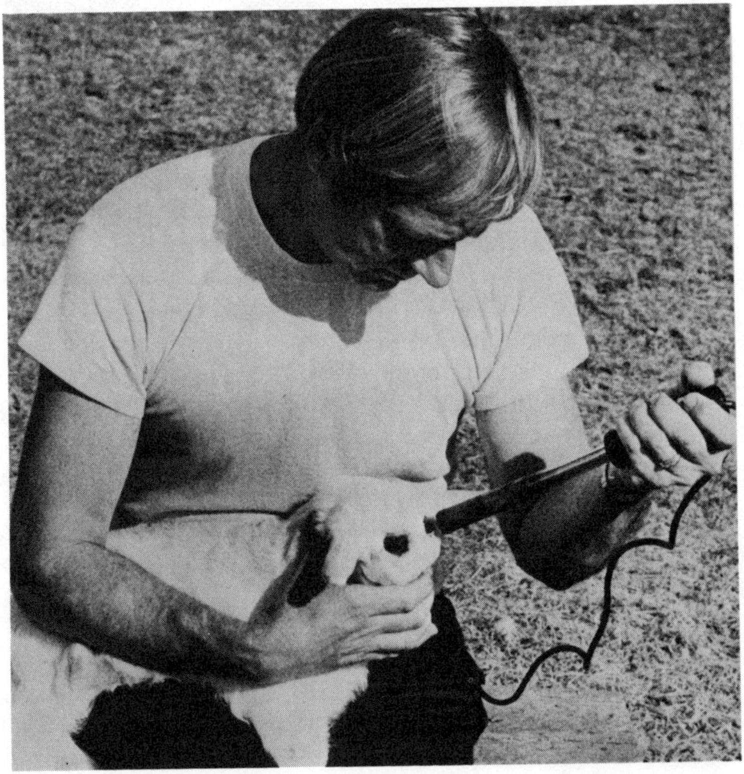

Fig. 7-11. Hold the kid securely while pressing the iron down over the horn button.

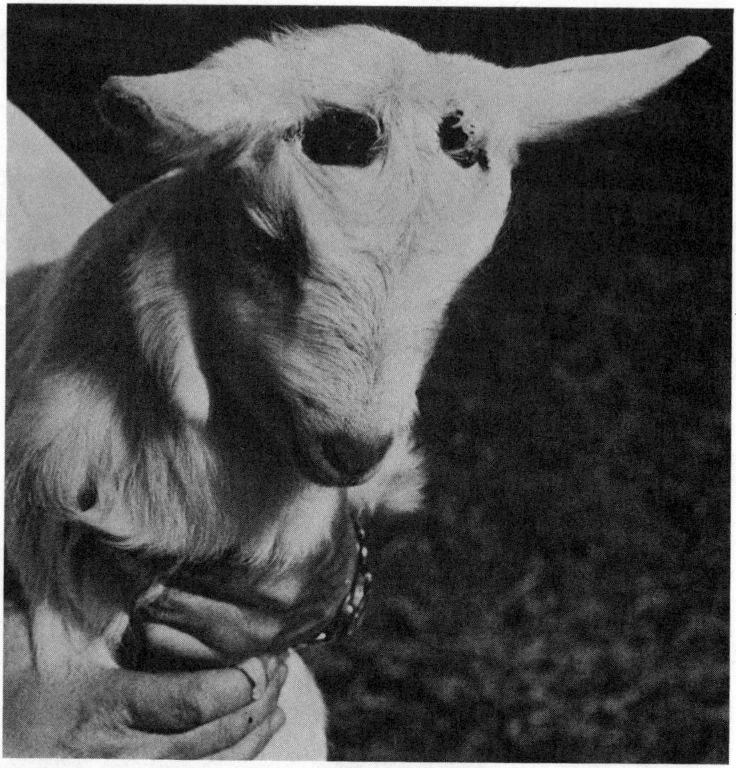

Fig. 7-12. Burn a circle around the base of each horn button.

so that the circumference of the iron tip burns a circle around the base of the button (Fig. 7-12). If you leave a hot iron at one point on the skull too long, you can kill the kid, so work carefully. The maximum time may depend on the ultimate heating temperature of the iron you buy, so read directions carefully. Although complications are unlikely with cauterized wounds, some dehorners like to apply a layer of carbolated Vaseline petroleum jelly to each treated button area.

Descenting a Buck Kid

Bucks have scent glands in the head that render them unwelcome in any gathering but that of female goats. If you have a male kid that you intend to keep as a buck, descenting may be required. There is no need to descent kids that you intend to castrate, because their scent glands won't develop with no testicles. Removing the scent glands has no effect on the buck's virility.

The scent glands are located just behind and slightly inward from the horn buds (Fig.7-13). In young kids the glands are quite small and easily burned out by a hot iron. If you can't locate the glands by feel, clip or shave the hair around the horn area. When you have located the glands. which are probably no bigger than a pencil eraser, burn them with a cherry red iron.

Some doe kids have very small bits of odor tissue in this location. Some goat owners claim that Toggenburg goats are most likely to have scent glands in the females, but others mention the Alpine breed. When disbudding your doelings, burn off the area just behind the horn buttons. This will insure that you won't have a doe that produces an odor, which can affect milk quality.

Dehorning Mature Goats

Dehorning a mature goat is a dangerous procedure and isn't recommended for novices. Either saw off the horns or use rubber bands of the kind designed for castrating calves.

The most satisfactory way to dehorn by sawing is to use special wire coated with a cutting material. You may have to get the wire

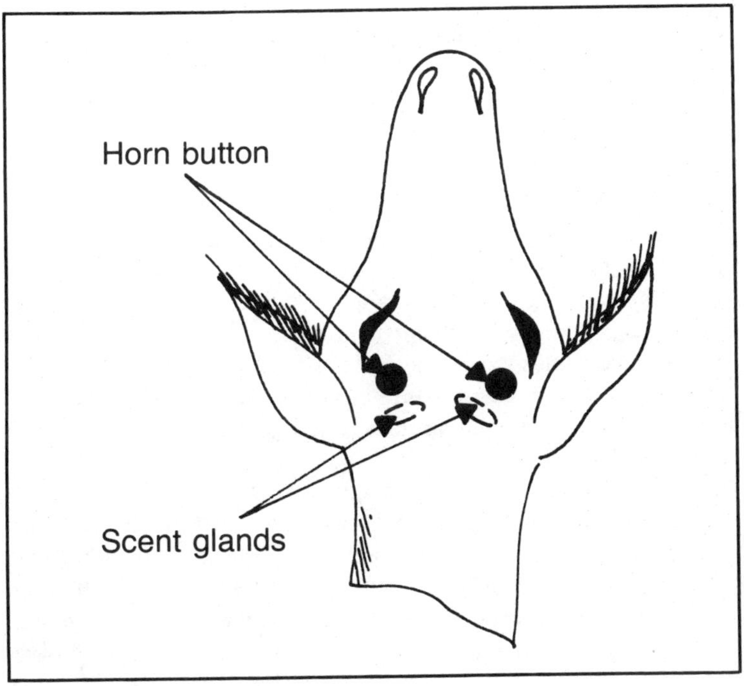

Fig. 7-13. Location of scent glands on a buck kid.

from a veterinarian or order it from a mail order supplier. You can make up a horn saw by installing short pieces of 1″ wood dowels at each end of a 24″ section of wire. One person holds the goat's head by placing one hand under the muzzle. The saw wielder stands at the front of the animal and pulls the wire back and forth so that it cuts off the horn at the base, from back to front. The wire cuts close to the skull, shaping the horn stub to match the contours of the head rather than leaving a scur. The hand under the muzzle holds the head up and prevents the wire saw from pulling the head down.

Dehorning an adult goat is probably a job for a veterinarian. Don't be surprised if your vet refuses the job.

Some goat keepers use rubber bands to eliminate horns. They get bands used in castrating calves and roll them onto the base of the horns. A notch is made at the base of the horn to hold the band. After a week or two, the theory is that the horns will drop off with no bleeding or other problems. It often doesn't work that way. One man I know tried it with one of his does. About a week after he had applied the bands, the doe hit her horn against a stanchion as he was getting ready to milk her. The horn fell off, but she cried so pitifully that he swore he would never use bands again. Her milk production dropped significantly. The man used a saw for his next goat dehorning task. That time he evidently hit an artery, because a fountain of blood erupted with each pulse beat. His motto now is to disbud kids and leave adults as they are. That is probably good advice for most goat owners.

CASTRATION

Only the best bred young males should be used for breeding. Male kids that are not going to be used as bucks should be routinely castrated when they are about a week old. If left uncastrated, they can't be safely kept with young does. Goats are sexually precocious animals, and a buck kid can begin to breed females when only a few months old. Owners who aren't aware of this possibility have had doelings stunted by coming in heat and getting bred in the spring after they were born.

I know of one case in which a doeling was born during January. She was given the name Jan to distinguish her from later does, and yet she had a kid in July of the same year. It was a male kid named Jules because of the July birthday. Jules was probably premature. The normal gestation period is 150 days. If he had been born on July 28, the breeding date would have been about March 1 for a full-term pregnancy. That would make Jan under two months of age at breeding. Many goat keepers have been skeptical of this story, but I know it happened because Jan was my goat.

One clue that Jules may have been premature was that the birth was totally unexpected. Who expects a kid from a six-month-old doe? Jan came into the shed on a hot day. When I saw her at the shed, I assumed that she was sick and told members of my family to remind me to check her in an hour. She had a kid in an hour. The sad part of this "accident" was that Jan grew to be only a little more than half the size of her dam, who was one of the best does I had. Unfortunately I had sold the dam to a fellow who needed goat milk for a baby who couldn't tolerate cow milk. I thought I had a replacement for the dam in Jan. Castrate your male kids unless you have separate facilities for them. Don't allow any bucks to roam with your young doelings. Most does don't come in heat during the summer unless artificially prepared for it, but there are exceptions.

There are three methods of castrating male kids—surgery, clamping, and banding. Many castrators have had bad experiences with banding, which causes pain over a long period of time, so that procedure will be omitted here.

Clamping

Clamping involves no bleeding, so flies or other pests are no problem. There is less danger of tetanus or other infections. Disadvantages are the expense of an emasculator, the chance that the job won't be done right, and complications with careless use. Because clamping is not 100 percent effective when done carelessly and the chances of complications for surgical castration are almost nil, surgery seems the most practical method.

If you want to try clamping, you must buy a Burdizzo emasculator or clamp. Get one designed for lambs; the sizes used for cattle will be too large. The object of clamping is to crush each cord that leads to a testicle, one at a time, to deprive the testicle of its blood supply. Take care to see that the clamp closes on a cord above the testicle. The clamp handles are closed all the way and held for about 30 seconds. When applying the clamp, make sure it doesn't extend far enough over the scrotum to include the center of the scrotum—just enough to enclose the cord. If the cord isn't fully clamped, the testicle may continue to receive its blood supply, or a partial supply, which can nullify the job.

Surgery

Castration by surgery should be done before the kid is more

than 10 days old. I prefer to castrate all unwanted male kids at one week of age. There will be few problems at this age. The kid will usually act perfectly normal within hours, or a day at most. In one case I let a kid escape castration until it was perhaps a month or two old. When I tried to pull out the cord, it seemed so well attached that I was worried about causing internal injuries. I cut the cord, which was about the size of a pencil. The kid bled so profusely that I was sure it would die. The kid became weak from bleeding and fell on its side in the goat shed. Evidently its blood pressure then dropped or something happened, because the bleeding stopped and the kid

Fig. 7-14. A kid held in position for surgical castration.

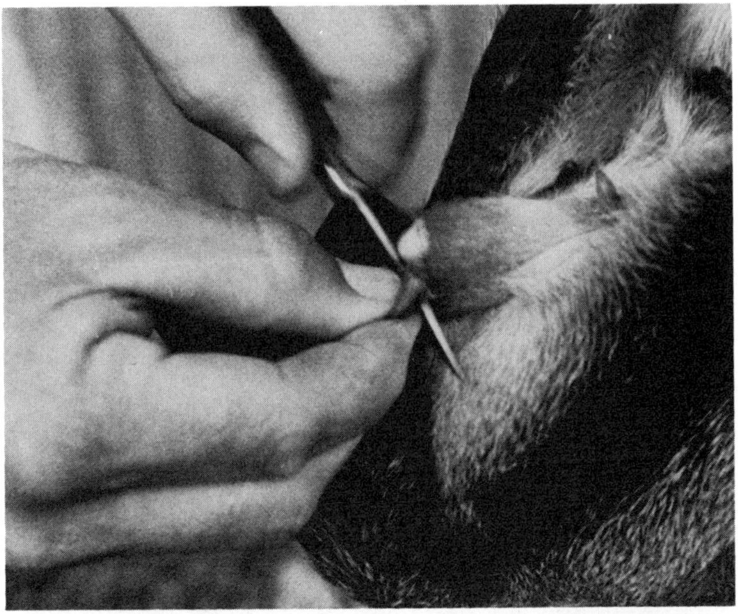

Fig. 7-15. Hold the end of the scrotum with one hand while cutting off about ½″ of it without cutting the testicles.

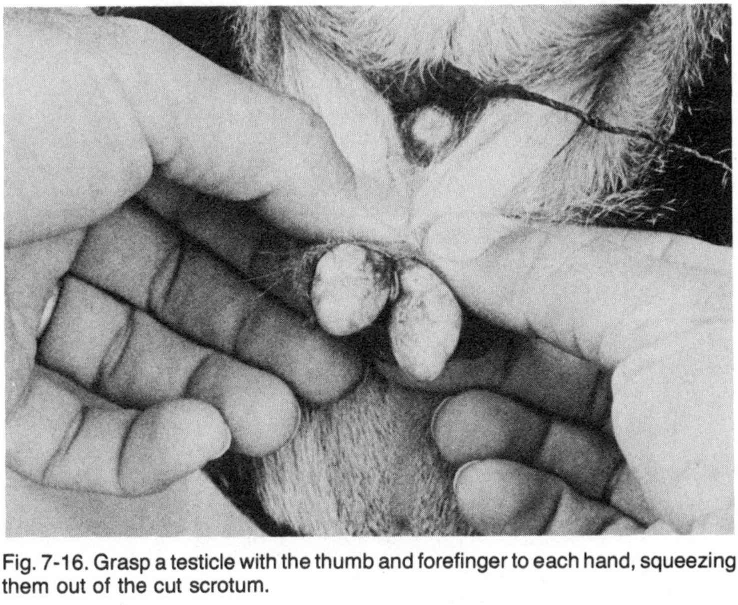

Fig. 7-16. Grasp a testicle with the thumb and forefinger to each hand, squeezing them out of the cut scrotum.

Fig. 7-17. Hold one testicle while you pull out the other one.

gradually recovered. I then decided to castrate at seven days. There will be little or no bleeding with baby kids, because you can pull the cord out rather than cut it.

The job should be done in sunny weather rather than damp weather. It's nice to get the job done before the fly season, but that depends on when your kids were born. I haven't noticed any fly problems for surgery that is done right. Use a sharp, clean knife disinfected with a 70 percent solution of alcohol or hydrogen peroxide.

To perform the surgery, have someone hold the kid's back against his body by grasping its rear legs and pulling them up

(Fig.7-14). Grasp the end of the scrotum with one hand and pull it away from the kid's body. This stretching movement will extend the scrotum, so you can cut off the end without cutting the testicles (Fig. 7-15). Cut off about ½″ of the scrotum's end for access to the testicles. Place the knife in a safe place and, using both hands, squeeze your thumbs and forefingers against the kid's belly. Bring them together so that you can grasp a testicle with each hand (Fig.7-16). Pull out the testicles one at a time, holding the other testicle firmly in place as you pull out the first one. As you pull out the testicle, the cord will pull out with it and finally detach from its interior connections (Fig. 7-17). By using both hands, you will prevent one testicle from being pushed up into the body cavity while you work on the other. After pulling out the first testicle, do the second. Use both hands if necessary to insure that the testicle doesn't slip away from you. After pulling out both testicles, examine the scrotum to see if any cord material is left hanging. If so grasp it and pull it out. The cleaner the operation, the better. If there is little or no bleeding, disinfection is usually unnecessary.

When you put the kid down, it will arch its back for a few minutes and walk with a stiff gait. It will be running and playing with the others the next day.

Chapter 8

Goat Health

One popular encyclopedia describes goats as the hardiest animals in the world. If a herd receives proper management, that statement is probably true, but with careless management domestic goats are susceptible to various health problems. Wild goats probably have fewer diseases than their restrained cousins. They can move freely from one area to another, so parasites and disease organisms don't build up in the soil around them.

BENEFITS OF BENIGN NEGLECT

The most important rule in planning a health program for a herd of goats is to avoid overtreatment. It may be fun to play veterinarian, but the purpose of any good management practice should be to improve the health of the goat rather than the skill of the practitioner. Sometimes animal owners feel that something must be done for an animal even if they have no idea of what is wrong or what medication is required. The prudent procedure is to let nature take its course. You can easily complicate matters by giving a treatment just to do something. It is irrelevant if the treatment makes you feel better. Goats that get sick often recover with no help from anyone, so don't tip the odds against them and rob them of their normal chance for survival.

Recent surveys have shown that many people believe they can follow any kind of life-style they choose, even if it is unhealthful, and then depend on the medical profession to work miracle cures when their bad habits catch up with them. This attitude is unrealistic for people. It would be unrealistic to keep goats in filthy quarters, give them contaminated hay or grain, and expect to make everything right with antibiotics. Prevention is much more effective than cures with goats and humans. Proper goat management pays off in reduced veterinarian costs and more productive animals. You can forget many of the ailments that infest a poorly managed herd with good management practices. The fewer medications you give your animals beyond those absolutely necessary, the better. When any of your goats become ill, remember that what you don't give a sick goat won't make it any sicker.

GOOD MANAGEMENT PRACTICES

Most desirable management practices for dairy goats are mentioned in other parts of this book, but some cannot be overemphasized. Provide the goats with clean, dry housing and proper food and water. Don't force your herd to share quarters with dogs or cats (Fig. 8-1). Contamination of feeds by either natural contaminants or agricultural chemicals should be avoided. Don't force goats to graze damp areas with no chance to reach dry ground. If any pasture area is small, move the goats to another site at regular intervals. Goats need shade, but protect your trees (Fig. 8-2).

One of the most important steps in maintaining a healthy herd is to raise your own replacement does instead of buying new ones. Each time you bring in a strange animal, you may introduce some

Fig. 8-1. Don't force your goats to share quarters with dogs or cats.

Fig. 8-2. Goats are natural browsers, and tree materials are good for rumen development.

infection into your herd. You can reduce the chances of importing a disease by carefully inspecting any new animals that you acquire, but no ordinary inspection can be 100 percent perfect. One man who has raised dairy goats for more than 20 years recalls that for about 10 years he never lost a goat to illness or even had one that acted sick. He decided to upgrade his herd and began buying and selling goats. He got some goats that for a while gave him more production. With a herd average of only two or three does, he lost five goats to illnesses in two years. Some deaths may have happened during that limited time by coincidence, but surely not all. He now believes that he imported a disease with at least one of the new goats. He put a premium on type with not enough consideration to the health of each doe.

CONTAGIOUS DISEASES OF GOATS

Although benign neglect can be used with many goat ailments, it should not be practiced when the disease is a contagious one, even if the only practical treatment is to dispose of the infected animal. If a contagious disease is ignored, it will spread through the herd. Some diseases carried by goats can infect their owners.

Abscesses

The goal is to maintain an abscess-free herd by refusing to buy an infected animal. Unfortunately young goats can have the disease in an early stage and show no symptoms. The first indication is a small lump often located near the throat, the shoulder blades, or in the flank. Isolate the infected animal from the rest of your herd at the first symptom. If it isn't a valuable individual, you might consider destroying it before the disease becomes endemic. The problem with this disease is that the soil can be contaminated and young goats will contract the disease, but they will give no evidence of it until they are adults.

If you decide to treat the animal, be prepared for a long siege of minor surgery plus antibiotic shots. Clip the hair from around the lump and wash the area with soap and water, then rinse and dry. Wear rubber or plastic gloves while doing surgery and treating the wound, then sterilize or burn them. To lance the *abscess*, make a vertical cut that reaches to the bottom of the lump so gravity will aid in draining the pus from the incision. Collect the pus with a paper towel or similar material, then burn it. Flush the wound and the area around it with an antiseptic solution such as tincture of iodine or hydrogen peroxide. Administer an antibiotic by intramuscular injection. Follow the directions that come with the antibiotic as to amount and frequency of treatment. If the lump is under the ear, you should probably have a veterinarian make the incision. This area contains important blood vessels, nerves, and salivary glands that could be damaged by unskillful cuts.

You may find that the treated animal develops new lumps. There likely will be internal abscesses that can interfere with bodily functions and cause death.

If you have a doe with abscesses, isolate newborn kids from her. Kids must have colostrum milk, right after they are born, but after a couple of days they should be changed to milk from a healthy doe or to milk replacer.

The best way to have an abscess-free herd is to buy only healthy animals and then, if a latent infection develops in one, promptly dispose of that goat. This may seem ruthless, but it can prevent a long quarantine program that can be expensive in time and perhaps cost. This is one disease where doing nothing is the worst choice. If you wait until a lump ruptures, the infection will spread around your pens and shed. This can cause further problems with goats that would otherwise be healthy.

Infectious Scours

Scours are often caused by Coccidia, and the illness is then called *coccidiosis*. This is usually considered a disease of chickens, but the organisms of this genus that infect goats are a different species from those harbored by chickens. The usual symptom is diarrhea or scours. It occurs mainly in young animals that can be stunted if the disease persists. The best treatment is prevention by sanitary practices and by isolating the kids from the adult goats. If kids are kept in confined quarters, they are more likely to get scours than if permitted to roam a large area. Some cases of scours have been cured by moving the animal to a new environment where some exercise space was available.

One way to treat scours is to use a diarrhea preparation intended for human use. Rexall drug stores usually carry an effective preparation called *Pabizol* that contains paregoric. Kaodene with codeine, kaolin, pectin, and other ingredients is another possible choice. There is always *Kaopectate.* Get some calf scour medicine for severe cases. This is usually administered to calves in the form of a *bolus* or large pill. Check the dosage instructions carefully. There will probably be a bolus/calf weight ratio that can be followed for goats. If the instructions recommend a bolus for every 100 pounds of body weight and your sick goat weighs about 30 pounds, you would divide each bolus into three parts and give one piece at a time. Be careful not to overmedicate a young goat. To give the bolus, hold it in the goat's mouth and run your fingers under its throat to try to induce swallowing. If you hold the bolus far enough back in the throat, the goat will swallow it.

Many goat owners dislike the lack of precision when they chop a bolus into pieces. They prefer to dissolve the bolus in water, then administer the solution with a large syringe. Goats swallow liquids much more readily than dry pills. Some preparations are almost insoluble. These materials can be ground into powder and made into water suspensions. Shake well before using.

For confirmed cases of coccidiosis there is a useful water soluble coccidiostat, *amprolium*, that Merck & Co. offers as Product 23091, *Corid*. Other companies produce similar preparations. Corid is usually sold in 4-ounce packages of 20 percent concentration. It is used either as a treatment for diseased animals, when mixed with 50 gallons of water, or as a preventative measure for exposed animals, when mixed with 100 gallons of water. Such volumes are useful for a large herd of cattle. If you want to dose a few goats, you can use 1 teaspoonful to the gallon for treatment and 1 teaspoonful to 2 gallons for prevention. Give the solution as the only source of water. During hot weather when most outbreaks of coccidiosis occur, goats will usually drink enough water for proper dosage. This treatment has the great virtue of easy administration. In theory it should be used only if coccidiosis has been diagnosed by a microscopic examination of the feces. As a practical matter the treatment is cheaper than taking your goat to the vet and is worth a try for any goat with summer scours. Don't use the milk or meat of any animal while it is under treatment.

If you are handy with a syringe, substitute a general antibiotic by intramuscular injection. Give proper attention to dosage with respect to the animal's body weight.

One precaution that helps to prevent scours in baby kids is to make sure they get enough colostrum milk. Many goat owners freeze surplus colostrum when it is available, so they can substitute it for the dam's milk if she dies or fails to produce enough milk for the kids. Some dairy authorities advocate feeding soured colostrum to young calves and baby goats to prevent scours. To sour the colostrum, milk out the udders of newly freshened does and store the milk at about 70 degrees Fahrenheit (20 degrees Celsius). After about two days it can be fed to kids by diluting it with 1 part warm water to 2 parts colostrum. In feeding milk to kids the milk should be kept near body temperature—about 100 degrees Fahrenheit or 38 degrees Celsius. Goat body temperature is about 103 degrees Fahrenheit or 39 degrees Celsius. Never try to sour colostrum from animals that are undergoing antibiotic treatment. The antibiotics will kill the desirable organisms that produce the proper kind of soured colostrum. Maintain clean utensils for colostrum that is saved for freezing or souring. Feeding kids milk that is contaminated or fed when at a temperature significantly below body temperature is a sure way to induce scours.

Some goat owners start kids drinking from a pan of milk soon after birth. This practice carries the risk that the kid will not stretch his neck to drink, directing the fluid into the rumen instead of the omasum where it should go. Milk in the rumen can cause gas and scours.

Mastitis

Mastitis is a disease of the mammary gland tissue in the udder. The organisms that cause mastitis can gain entry through the teat orifices, so cleanliness while milking is an important measure in disease prevention. The condition that makes goats most susceptible is an injury to udder or teats. To reduce the chances of your doe getting mastitis, remove any unnecessary barriers in the pens or pasture, especially those that extend to normal udder height. Goats might be tempted to jump over or pass through them.

In addition to washing the teats and udder at milking time, use a disinfectant solution. Mix ½ cup of household bleach with 4 gallons of water or about 8 cc of bleach per liter of water. Use a fresh paper towel for each doe to reduce chances of contamination.

Choose does with the kind of udders that are less susceptible to mastitis. Pendulous udders that almost drag to the ground are more likely to become infected than are compact udders with wide attachments.

Loafing areas should be dry and clean. Bedding material should be changed frequently. Clean, bright straw makes good bedding material. Wood chips will serve well if available in your area.

Sometimes milking practices can increase the risk of mastitis. Each session should be as calm and undisturbed as possible, particularly for young does. If the doe becomes nervous or frightened by loud noises, actions of other goats, or by the milker's bad temper, she will involuntarily hold up her milk. This is a normal reaction produced by her adrenal glands. If conditions are wrong, she is incapable of controlling it. The udder can't be emptied of milk, and that predisposes the udder to problems such as mastitis. Use a dairy cow teat dip solution after milking. Put hay into the hay feeder immediately. This induces the doe to stand at the feeder rather than to lie down. Her teat orifices are slightly open after milking, making them subject to infection at that time.

The typical mastitis symptoms are lumpy or hard udders that produce stringy or clotted milk which is difficult to draw from the teats. Blood in the milk is not an indication of mastitis. Sometimes the bleeding is so slight that it is only visible as a pink layer near the bottom of a vessel after the milk has stood for hours. This symptom is probably the sign of a small blood vessel break rather than mastitis.

If a mastitis infection becomes acute, the udder can be swollen and feverish with obvious pain. The infection may affect only one side or both sides of the udder. The milk can be watery and yellow as well as clotted.

If you suspect mastitis, verify the diagnosis by either taking a milk sample to your local veterinarian or by obtaining a test kit and doing the test yourself. The confirmation or even suspicion of the disease suggests several measures to keep it from spreading. Nursing kids should be removed from the doe, because they can spread the disease from one teat to the other. The doe should be isolated from other milking does. To further prevent spreading the disease, milk the infected doe last.

Properly managed goat herds rarely show mastitis infections, but when they do, treatment is difficult. Your veterinarian can recommend the proper antibiotic infusion preparation. This is inserted into the teat orifice. The antibiotic is delivered into the udder. Follow instructions carefully. You will normally use this treatment for three days. You must then avoid using the milk for a longer period, usually about a week or more after treatment. The antibiotic can produce allergic reactions in many humans and should not be used by anyone.

The usual procedure for infusing the antibiotic into the udder is to first milk out the udder completely. If milking is made difficult by the infection, a towel dipped in warm water and used to massage the udder will often aid the process. Wash the udder with soap and water. Dry it carefully. Make sure that the teats are especially clean. Wash your hands carefully. Swab the teat ends with a 70 percent solution of rubbing alcohol and clean the antibiotic container with alcohol. Use about one-half the usual dose required for a dairy cow in the two halves of a goat's udder. Massage the udder after infusing the medicine into it to aid in getting the antibiotic into the infected area. Continue treatment daily until the udder becomes normal to the touch.

If it is impossible to milk out the udder because of the infection, a general antibiotic, via intramuscular injection, should be tried for a few days before inserting a local antibiotic into the udder. Some veterinarians prefer to use a general antibiotic rather than local infusion in most cases. They say that infusion can introduce new bacteria.

Milk from a goat infected with mastitis should never be used unless it is pasteurized. This milk will contain some of the organisms that produce the mastitis. If left standing at room temperature for even a short time, the milk can become loaded with bacteria.

Pinkeye

Goats that are kept in a windy, dusty environment are more susceptible to *pinkeye* than those protected from flying dust or other particles. Pinkeye can be transmitted by goats that show no symptoms. The first symptom is likely to be a watering or clear discharge of the infected eye. The eyelid may droop. These same symptoms can be produced by exposure to various irritants. If a cloudy area appears on the cornea, that is a positive indication of pinkeye.

Isolate any obviously infected animals for a week. Obtain an antibiotic from your veterinary medicine supplier and apply it to the infected eye. These ophthalmic antibiotics are available in spray cans, which makes for easy treatment. Continue the treatment until the eye returns to normal appearance. Consult a veterinarian for stubborn cases.

One measure that helps prevent pinkeye is to install hay racks at the proper level. If an elevated hay feeder is used, one that requires the goats to reach upward to pull out the hay, dust particles will rain down on them. This kind of situation will predispose the herd to eye problems.

Foot Rot

This is an infectious disease but can be kept under control in most herds by proper attention to hoof trimming. If you set up a regular schedule for this job and keep your goats' hooves neatly trimmed, following the procedure described later in this chapter, you should have little trouble with *foot rot*.

Another protective measure is to provide a well drained site for pens and sheds. Goats that are kept in damp, filthy quarters are more susceptible to foot rot than those living in a dry, clean environment. Sites with rocky soil are better for foot health than overly rich bottom land. Goats do well on soils that are poor for growing agricultural crops.

The first symptom of foot rot is often a lameness in the infected foot. Closer inspection will show a dark area between the hoof shell and the sole of the foot. If you trim the hoof, there will be a strong, offensive odor indicating bacterial growth.

To treat the condition, isolate the infected animal and trim the hoof. Remove all the infected tissue, even if it causes bleeding. Dip the foot in a solution of copper sulfate. To make up the solution, dissolve 1 pound of bluestone crystals (copper sulfate) into 6 gallons of water. In metric measure dissolve 20 grams of copper sulfate into 1 liter of water. You can make up any amount of this solution that is convenient; just scale the quantity of bluestone and water up or down to maintain the right concentration. A quarter of a pound of bluestone in 1-½ gallons of water would work well. Use this solution to treat all your goats' hooves after each trimming, especially if you have had any incidence of foot rot in the herd.

Foot rot in the forefeet is bad enough, but the animal can usually continue to graze on its knees. If the rear feet are infected, the animal will often just lie down and be unable to move about. An intramuscular injection of antibiotic is helpful in advanced cases.

Pneumonia

This disease usually occurs only where goats are overcrowded or kept in a poorly ventilated shed during cold weather. If the shed is partially heated, the problem is more likely to occur. Extreme dust concentration is a contributing factor, along with sudden changes in weather or housing conditions. *Pneumonia* is another disease where proper management reduces the chance for illness. The problem is complicated if the infected goats also have lungworms.

The symptoms of pneumonia are rapid breathing, a nasal discharge, and fever. A sick goat should be isolated and treated with antibiotics, but the animal may never be the same again. This is a problem that usually requires a veterinarian.

Enterotoxemia

Enterotoxemia is a disease that is dangerous to your goats even with good management. The herd can be vaccinated against it, but vaccination alone is not enough. This bacterial disease is precipitated by poor feeding practice, and no vaccination will make up for careless management.

Another name for enterotoxemia is overeating disease. This is a fairly descriptive term, but feed switch disease might be a better choice of words. Symptoms are most likely to occur after a sudden change in feeding habits. Here are two case histories to illustrate how carelessness can predispose your goats to the illness (Fig. 8-3).

Case One. A two-year-old doe was purchased on a weekend, moved to her new home, put on a lush grass pasture, and fed a

Fig. 8-3. Never leave concentrates where goats can get to them.

concentrate that consisted of 50 percent grain and 50 percent pellets, which included molasses in the formula. On Tuesday the owner bought a new batch of feed, which appeared to be more moist than usual. The moist look suggested that it was fresher and perhaps had a higher concentration of molasses. On Friday evening the doe refused to eat, but she otherwise appeared normal. She refused feed on Saturday morning, so the owner kept her in a pen for observation. When he checked her at noon, she had begun to show diarrhea.

At 2:00 P.M. he gave the doe a dose (3 cc) of a diarrhea medicine called Kaodene that he had purchased at a drugstore. It contains codeine, kaolin, and pectin among other ingredients. If you use such over-the-counter preparations for diarrhea cases, ask your local pharmacist for advice on what is available. If you live in a rural area, it is likely that farmers or ranchers have used these nonprescription items and reported their effectiveness to the pharmacist. In this case the medicine, or just the passage of time, appeared to take some effect by Saturday evening. The doe appeared to be slightly more alert and ate some hay. This was her first hunger sign in more than 24 hours. The owner repeated the dosage of diarrhea medicine. The doe took some hay the next morning, but she still showed evidence of scours. Another dose of Kaodene was given. By Sunday evening the doe was alert and eager to eat.

The owner gave a final dose of the medicine and then called the previous owner to get clues to the problem. He learned that the doe was used to a straight grain ration and a fairly sparse pasture on dry hilly ground. He switched her to oats, but gave her only a limited amount for the next few days. She was fed all the hay she would eat, then released to the pasture for a short time in the afternoon. Gradually he increased her feed allowance and her time in the pasture each day. Later he supplemented the all-oat diet with a protein-rich calf feed. The doe was gradually switched to the same feed used for the rest of the herd. There were no more problems with scours.

Case Two. A 1½-year-old doe was purchased when about 4 months old. She was bred at 8 months and had kids at 13 months. She was extremely nervous from the time she was moved to her new home. The doe refused to be tamed by normal methods. She often failed to eat well if bothered by an older doe. Her owner failed to latch the feed room door securely on June 25, and the goats got into the room. When the oversight was discovered the next morning, it appeared that the goats had gorged themselves on a 50 percent corn-50 percent pellet feed. They showed no obvious effects. Dosing with ½ cup of vegetable oil may be helpful for incidents such as this one.

The owner went on vacation for a few weeks and left the care of the goats to some friends. When the owner returned on July 27, he noticed that the young doe looked thin, as if she had eaten poorly while he was gone. The explanation was that to simplify management, the goats had been fed in a common feeder. The nervous goat had failed to get her share. The owner put the young doe in a separate pen and gave her extra rations. The next day at 7:00 A.M. she refused to eat and began to cry as if in pain. At 8:00 A.M. she moved to the outside of the shed, crying constantly. At 9:00 A.M.she got up again, staggered a few steps, and fell. She gasped a few times and died.

At 10:00 A.M. an autopsy by a veterinarian showed that her internal organs were quite healthy except for the small intestine, where a slight irritation was found. The vet's verdict was enterotoxemia that came on so quickly there was little or no internal damage. A review of the case history, including the unlocked door incident, convinced him that the trouble had been caused by the first gorging and the final extra feeding. The first gorging had probably upset the symbiotic relationship of the organisms in the rumen and the goat. While still recovering from that sudden food availability, the goat had been hit again with another quick change. In the second case she was literally killed with kindness.

These case histories show what you should not do in managing goats. Never make a sudden change in feed. If you feel that a previous owner hasn't been supplying a goat with a ration high enough in protein, make the change to an adequate ration gradually. If your goat has been eating feed with a protein concentration of 12 percent, don't suddenly shift to an 18 percent ration and expect her to do better. This kind of change can upset the goat's digestive system and cause organisms in the system to produce a poison that can kill the goat.

Prompt use of antitoxin can bring temporary relief in acute cases like the second one. Take the goat to a veterinarian as soon as possible. If she shows sudden improvement after the treatment, that can be considered a positive diagnosis of enterotoxemia. The treatment should be repeated at four-hour intervals, then at longer intervals if necessary. Quick action can sometimes save a valuable animal, but it's probably a long shot. If you have valuable goats, consider having them vaccinated.

Paratuberculosis or Johne's Disease

This chronic disease can end in death after many months. *Paratuberculosis* is a disease of the upper digestive tract, so diarrhea may not be a symptom unless the animal is in the final stages of the disease. Infection occurs when feed or water is contaminated by the feces of sick animals. Individuals may show temporary improvement at times, but any stress such as kidding will bring on a relapse.

For positive diagnosis, check with a veterinarian on the availability of testing laboratories in your area. A sample of fecal matter must be taken from the rectum and sent to the lab for culture tests requiring many weeks. These laboratories are usually available at universities that offer degrees in veterinary science. Cull any animals that show a positive test.

DISEASES OF GOATS THAT CAN INFECT HUMANS

Some diseases can be spread to humans who drink the milk of infected goats. Such transmission is highly unlikely with proper care, so you should be aware of the possibility. The following diseases are most likely to be encountered.

Brucellosis

This disease is also known as Bang's disease, Malta fever, Texas fever, Rio Grande fever, goat fever, and undulent fever. It is not common in American dairy goats, but the disease's past history causes many people to suspect goats as helping to spread it.

Brucellosis got its name from Dr. David Bruce, a British physician who headed a team of doctors sent to Malta in the early twentieth century to find the source of a strange illness affecting British military personnel who were stationed there. Goats were the main source of milk in Malta. When it was discovered that milk from infected goats produced the fever in humans, goats became associated with Malta fever. Although the disease could have originated with goats, the first authenticated case in humans in the United States was reported in 1905, involving a nurse who had never used goat's milk.

About when the disease was discovered in a human in the United States, the United States Bureau of Animal Husbandry sent an agent to Malta to purchase dairy goats. The goats, 61 does and four bucks, were shipped from Malta on the S.S. Joshua Nicholson to Antwerp, where they were transferred to the S.S. St. Andrew bound for the United States. The crews of both ships used the milk produced by the goats during that shipment. There was a 12-man crew on the first ship, and eight men developed the fever from 18 to 34 days after the goats were placed on the ship. Those who didn't get sick had either boiled the milk they drank or refused it. The goats were placed in quarantine after arriving in the United States, but one woman who drank the milk from the goats became ill. All the imported goats were then slaughtered.

A few years later investigators reported that brucellosis was common in southwestern Texas and apparently had been for many years. This probably accounts for the names Texas fever and Rio Grande fever. Spanish explorers may have introduced the disease into that region by bringing goats from Europe.

The most notable symptom of brucellosis in animals is spontaneous abortion. A contagious abortion was noted in cattle in the southern United States about the time of the Civil War, but at that time the germ theory of disease hadn't been developed. A veterinarian of the time recommended that cows which aborted be isolated from the herd. His reason was that other cows would not see the animal abort and would not be inclined to abort in sympathy with the aborting one. You don't always have to fully understand the disease with which you are working to find helpful treatment if you are an alert observer.

Veterinarians usually advise that goats be tested for brucellosis before using their milk. This is debatable because of the disease's rarity in this country now, but the test isn't expensive and does give a sense of security when your goats pass it. The testing should be done before breeding a doe. Pregnancy sometimes causes a false positive reaction, and even a doe that has just kidded can give a false test. An animal that gives a persistent positive test should be slaughtered. By handling the raw flesh of an infected animal, you can get the disease just as easily as by drinking the raw milk.

Leptospirosis

This disease, commonly called "lepto," can be fatal to goats and causes high fever in people. It often produces signs of jaundice and kidney problems in humans. *Leptospirosis* is often spread by dogs or rats. It can be prevented by vaccination. Check with your veterinarian.

Listeriosis

Listeriosis is sometimes called circling disease because of the most noticeable symptom, which is circling, always in one direction. The first symptom is usually a fever and listlessness, followed by a paralysis on one side. This is noticeable by the drooping of the ear on that side, particularly if the goat is of a breed that normally has erect ears. The circling symptom comes next. Death usually occurs within a day of the first circling movements. This form of the disease is known as the *encephalitic* or brain fever form. In another form the chief symptom is abortion. It should be suspected if a doe that has been tested for brucellosis aborts. This disease can be transmitted to humans, where it occurs as meningitis or glandular fever.

Call a veterinarian if you suspect listeriosis. Confirmation requires lab testing. If one animal in your herd dies from this disease, you may be able to save others by observing them for early symptoms and treating them with large doses of penicillin, 1 million units or more, administered intramuscularly. The injections should be continued daily for three days.

Proper sanitation in feeding forage and concentrates will reduce the chances of a listeriosis outbreak.

Tuberculosis

According to recent surveys, *tuberculosis* is extremely rare in goats. Tuberculosis was formerly found in dairy cattle until an eradication program eliminated it in many areas. Milk from an infected animal can transmit the disease to humans unless the milk is pasteurized.

METABOLIC DISEASES OF GOATS

These diseases are sometimes caused by similar conditions to those involved in enterotoxemia, but the nature of the disease is different in that no bacteria or other organisms are responsible. The art of prevention is again worth much more than a cure in each of these diseases. The chief cause of metabolic diseases can be summed up in one word—carelessness (Fig. 8-3).

Bloat

Bloat is most likely to occur when a goat kept in a dry lot is suddenly turned out to a lush pasture. The problem is more severe if the pasture is a legume such as clover or alfalfa. The first symptom is a balloonlike appearance of the animal's sides, particularly the left side. The goat will often cry in pain and finally stagger and fall, frequently rolling over so that it lies on its back. Death usually occurs soon after this stage is reached.

Don't switch to a pasture that is richer and faster growing. If you are moving goats from a dry pasture or lot to one that has bloat-causing plants, make sure the animals get all the dry hay they will eat first. Leave them on the new pasture for a short time, then move them off for the rest of the day. By gradually introducing them to the "better" pasture, excessive bloat is less likely to occur. Even with care you will probably see the initial signs of bloat, as the goats return from the greener pasture with their sides sticking out.

If the bloat is serious, give the animal a cup of vegetable oil such as corn oil or other cooking oil. You can also use mineral oil sold in pharmacies. In severe cases of bloat the goat can not burp to relieve the condition. Look for regular movement in the goat's left side above the flank; this movement indicates belching and is evidence that the problem isn't acute. When a goat appears to be excessively bloated, keeping it on its feet and walking it about may help dislodge the gas.

Ketosis

Ketosis affects dairy goats just before or after kidding. It is brought on by stress, but can be caused by overfeeding in early pregnancy and underfeeding in late pregnancy. Some does probably have faulty metabolism, making them more susceptible to ketosis. It is often a disease caused by poor feeding. A doe that is fed properly and gets enough exercise will be in good condition to meet the demands of late pregnancy, kidding, and early lactation.

The first symptom is likely to be a loss of appetite. Be alert to any does that miss a feeding. The earlier you spot the symptoms, the better the chance of saving the doe. A doe with ketosis soon becomes listless and may stagger as she walks. If the case becomes acute, she will lapse into a coma and die. Smell the doe's breath to make a more positive identification of the disease. If there is a fruity odor, that is an almost sure clue that ketosis is the problem. To get a special ketosis test kit, check with your supplier of veterinary medicines.

When a doe misses a feeding, it's time to act. Try to get her to eat by tempting her with her favorite food. If it is midwinter, take something such as carrots or lettuce out of your refrigerator to get her started eating. Most goats like oats or crimped corn, so get some of each and see if she will eat. Mix ½ cup of molasses into 1 quart of lukewarm water and see if she will drink. The molasses will go directly to the omasum, where it will get into the blood more rapidly than food that goes to the rumen. Many goat owners feed propylene glycol for ketosis symptoms, giving a doe from 4 to 6 ounces at 12-hour intervals. If improvement doesn't occur soon, call your veterinarian.

Give each kidding doe the molasses and water treatment just after kidding. This improves the chances for a normal recovery from kidding.

Milk Fever

Milk fever strikes the best producers in a herd. It is caused by a deficiency of blood calcium. Dairy cows subjected to milk fever often die, but goats aren't usually so severely affected.

Although the disease is one of calcium deficiency, it isn't caused by calcium starvation. The opposite situation may be true. It can be compared to hypoglycemia in humans, a disease of low blood sugar often caused by eating too much sugar.

When a doe kids and begins lactation, she must draw on the calcium stored in her bones to furnish enough blood calcium for milk production. If the doe is fed a high calcium diet, she apparently does not activate the calcium utilization system in her body. When lactation time arrives, she is unable to cope with the sudden calcium demand. Some veterinarians suggest that no alfalfa hay or other hay rich in calcium be fed the last few weeks before kidding, but this is a controversial recommendation. You should examine the labels of any feeds used during this time. Limit calcium by either selecting a concentrate with no calcium additives or limiting the amount of alfalfa hay and substituting a grass hay for forage.

The symptoms that occur in goats are weakness in the rear legs, which results in unnatural use of those legs. The animal may be constipated. In severe cases the doe will get down and can't get up. Coma and death follow this stage. These symptoms will usually follow the time of kidding by a day or two. To distinguish between milk fever and ketosis, check the goat's breath or milk for the fruity odor that occurs with ketosis. Call your veterinarian if you suspect milk fever. He will inject calcium borogluconate into the bloodstream via a vein. This ordinarily is a veterinarian's job. Failure to treat the doe could result in further complications such as enterotoxemia or even death.

WORMS

Goats have a large area of internal digestive organs compared to total body size, which should make them susceptible to worms except for their fastidious eating habits. Wild goats probably have fewer worms than domesticated animals because they roam the countryside. Worm eggs don't build up in the soil. Closely confined animals placed in a damp place are most susceptible, along with goats that are fed on the ground (Fig. 8-4). Avoid heavy contamination of pastures or pens with eggs and larvae. This can be achieved by pasture rotation and keeping pens and sheds dry.

There is some debate about the advisability of a regular worming program. Some authorities claim that frequent worming deprives goats of the ability to build up a natural resistance to worms. Proponents or regular worming practice say that if you don't worm at intervals of six months or so, worm infestation can become so heavy in a goat that recovery will be difficult and slow. They also say it can predispose the animal to enterotoxemia or other diseases. A weakened goat, they say, cannot have the normal resistance to disease of a worm-free, healthy goat. Heavy infestation in an animal can cause it to become emaciated and in poor condition for any stress. There is some question as to whether such infestation is likely if good management practices are followed.

If you don't follow a regular worming schedule, check your animals to see if worming is necessary. The approved method is to collect a sample of feces from any goat that you suspect of having

Fig. 8-4. Don't put feed on the ground.

worms and take it to your veterinarian. A microscopic examination can determine if worm eggs are present.

Some goat owners used an alternate method of checking for worms. They periodically roll back the eyelid on each goat and look for signs of anemia. If the interior surfaces of the eyelids are bright pink, the goat passes the examination. If they are pale pink or almost colorless, the goat gets wormed. You must know the normal eyelid color of the goat to use this method. Some seem to develop a deeper hue than others, so you may think one goat has more worms than another when the difference is one of heredity. This symptom does not occur with every kind of parasite.

The most universal symptom is a goat may fail to thrive even with good feeding practice. Some worms cause goats to have diarrhea while others produce constipation. A heavy infestation can cause death. If a goat shows a sudden illness, don't react by giving it worm medicine until you have eliminated the possibility of other diseases. All worm medicines work because they are toxic to the worms. Any substance that is toxic to one kind of living thing is likely to be at least slightly toxic to others. A very sick goat can still recover if left to its own recuperative power. Giving toxicants to a sick animal, though, is something like kicking it when it's down.

The worming medicine most widely used for goats is probably *thiabendazole* or TBZ, which is sold as a paste or bolus. It is a broad-spectrum wormer; it works against a variety of internal parasites. Thiabendazole is also relatively safe for goats and probably your first choice for a worming medicine.

Goat owners who worry about lungworms usually set up a schedule of using TBZ and tramisol at alternate wormings. The typical time lapse between wormings is six months (Table 7-1). Tramisol gives some protection against lungworms.

One advantage of having a veterinarian check for worm eggs is that he can probably identify the kind of worm infestation in a goat and then use a wormer specific for that parasite. This is more expensive than doing your own worming, but consult a veterinarian if you have a stubborn case.

Make sure you read the instructions for any wormer. Don't use the milk for the period specified by the manufacturer. Some goat keepers use various wormers for dairy goats, including wormers designed for use with horses and other nondairy animals. In using these medicines you are using your goats and, to some extent, yourself as guinea pigs. Dairy animals sometimes retain certain substances in their body fat and pass them off gradually in the milk fat they produce. This makes the practice of using all kinds of wormers for dairy goats rather questionable. It may give someone a thrill to feel that he has discovered a new way to worm goats which is superior to the one followed by the veterinarian establishment, but the goal should be a healthy goat rather than kicks for the owner. One man told me he allowed a 30-day period for not using milk on medicines that haven't been cleared for dairy animals. This is guesswork.

FIRST AID FOR GOATS

You will rarely need to use your first aid kit for goats, unless you have permitted too many dangerous obstacles to accumulate in the goat pasture or shed. If you have fences with barbed wire at low points where goats can brush against them, you may have an occasional udder or teat injury. Little more than cleaning the area and swabbing with hydrogen peroxide is necessary for minor cuts or scratches.

One kind of injury that occurs with young kids, especially if you permit them to climb over wood piles or other similar mounds, is broken legs. Such injuries can be self correcting, but a splint aids in keeping the bones straight during the healing process. Goat's bones mend rapidly, so you probably won't have to use a splint for more than about 10 days.

You can use either a thin piece of wood (Fig. 8-5) or a section of thin-walled steel tubing, to splint a leg bone. Gas line tubing from an old car is ideal. When using tubing, bend it into a U-shape so that the goat's foot is at the bottom of the U. Veterinarians make a splint this way, orienting the tubing so it goes in front and to the rear of the leg. A piece of narrow wood is fine for many breaks. Wrap the splint and the leg with a strip of elastic bandage material or use cloth strips and pin the cloth. Don't make the wrapping so tight that the circulation of blood in the leg will be affected. Feel the foot to see if it is warm; coldness means that the wrapping is too tight.

The advantage of the U-shaped tubing is the goat can begin to use the leg sooner, using the bottom of the U for a foot. Don't worry that a goat will use the leg too much. A goat with three legs is almost as nimble as a goat with four legs.

POISONOUS PLANTS THAT ARE TOXIC TO GOATS

Goats are unlikely to be poisoned by plants under normal conditions. If a goat can freely choose among various plants, it will

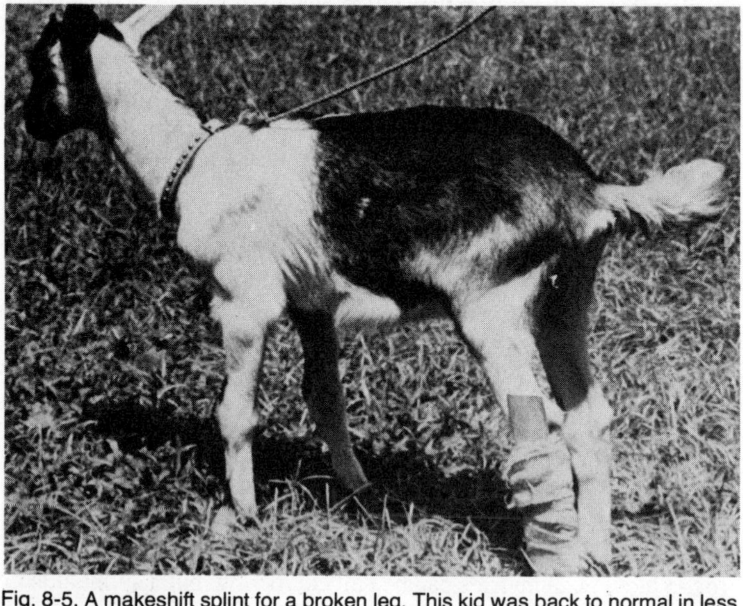

Fig. 8-5. A makeshift splint for a broken leg. This kid was back to normal in less than two weeks.

Fig. 8-6. Goats kept in a dry lot will eat almost any green plants that are available.

nibble here and grab a mouthful there, but it won't usually stop and gorge on a single plant. Goats sometimes eat dangerous plants if kept in a small area near them. If a goat has been kept in a dry lot for a long time and can suddenly reach through the fence to a poisonous species, it probably will (Fig. 8-6).

I know of one family who allowed a young doeling to have access to their lawn all last summer. *Yew* is one of the most poisonous of plants (Table 8-1). The doeling ignored two yews growing in a foundation planting at the house, although she ate other plants near them. Apparently the yews didn't attract her.

Another situation that induces goats to eat poisonous plants is sudden access to green plants after confinement to a dry lot, or even fresh access to a pasture containing poisonous plants after long restriction to another pasture. Goats like variety more than most animals, so the availability of a new plant will entice them even if they normally avoid that plant. The person who made up the old adage about the grass being greener on the other side of the fence was probably thinking about goats. Goats must be introduced to a new pasture with strange plants slowly, preferably after a full feeding of hay or other roughage. Toxic wood plants are especially dangerous to goats kept in a dry lot or grass pasture for a time.

If goats are given an adequate diet of safe plant material, they are unlikely to eat enough poisonous plants to be seriously affected. Again, one reason for goats' ability to dispose of some poisonous

Table 8-1. Some Common Poisonous Plants.

Plant	Poisonous Part	Special Conditions
Arum	All parts	
Belladonna	All parts	
Braken fern	Foliage	
Castor bean	Seeds deadly	
Deadly nightshade	Green berries, wilted leaves	
Elderberry	Foliage	Drought
Foxglove	Foliage	
Hemlock	All parts	
Henbane	Juice	
Jimsonweed	All parts	
Laburnum	Seeds	
Lamb's-quarter	Foliage	Drought
Larkspur	Foliage	
Locoweed	Foliage	
Monkshood	All parts	
Oleander	All parts	
Pigweed	Foliage	Drought
Rhododendron	Foliage	
Rhubarb	Foliage	
Smartweed	Foliage	Drought
Sorghum	All parts	Drought
Sudan	All parts	Drought
Water hemlock	All parts	
Wild cherry	Wilted foliage	
Yew	Foliage & fruit	

materials with no effect is they like variety. They will take a few bites of the poisonous plant, then move on to other forage.

TYPES OF POISONOUS PLANTS

Some plants are poisonous at all times. Others are poisonous only under certain conditions. Goats can eat an unlimited amount of green wild cherry leaves with no effect. If a branch is broken from the tree and the leaves are slightly wilted, a few leaves can cause death. The newly wilted leaves contain hydrogen cyanide, which later dissipates and makes the dry leaves safe again. The typical wild cherry poisoning occurs after a storm when a limb is blown down. Goats or cattle often gather under the tree to eat the leaves. If the leaves are wilted, they die there. In earlier times many an insurance claim was paid for animals lying dead under a wild cherry tree in what was assumed to be a lightning strike, but it was really hydrogen cyanide poisoning. Peaches, plums, cherries, and apricots can cause the same poisoning if goats have access to them after a limb or even a twig is broken off. Cultivated fruit trees cause no problem for goats, because no one is likely to provide the goats with access to them.

Another class of plants that is poisonous only under certain conditions are the tall grasslike plants of the *Sudan* and *Sorghum* group. When these plants are grown under conditions such as drought that stunt them, they contain enough cyanide to cause death in farm animals. The danger is greatest when the plants are small during the drought.

Some plants that are safe during normal climatic conditions accumulate a concentration of nitrates during dry weather that is too high for safety. Some common weeds such as *pigweed* and *lamb's-quarter* that goats like can become poisonous during a drought. One woody plant, *elderberry*, is noted for the same reason. Oat hay produced under cold wet conditions also accumulates nitrates. Nitrate poisoning can cause abortion and even death of the doe.

Some of the most dangerous plants are those that are poisonous regardless of climate or growing conditions. These include plants in the nightshade family, which have green fruit that is poisonous until it ripens, *jimsonweed, foxglove, locoweed*, and others. The botanical name for the foxglove genus is *digitalis*, which is the source of the powerful heart stimulant sometimes prescribed for heart patients. Humans sometimes mistake foxglove for wild *comfrey*, which is often used as a herbal tea. In one recent incident a couple in Washington state made tea from foxglove, and both died because of disturbed heart rhythms. The same mistake, taking foxglove for comfrey, could be made by goat keepers who cut forage and bring it to goats.

If you live in the western United States, you have probably heard stories about cattle eating locoweed. Apparently this plant is addictive, causing animals to get so hooked on it that they choose it in preference to other plants. When they ingest enough, the "loco" symptoms occur, then death.

In addition to the wild plants that goats find in many wooded areas, many horticultural plants, especially ornamentals, are poisonous. Your goats will unlikely have access to these plants, unless you permit them to roam your lawn or stake them on a tether near your house (Fig. 8-7). Make sure you know what kind of plants you have and if they are dangerous. Table 8-1 lists some of the most poisonous plants, but it isn't complete. There are literally hundreds of poisonous species.

There is one kind of poisoning that affects some breeds of goats more than others. It is called *photosensitivity*—sensitive to light. Light-colored animals such as Saanens are subject to photosensitivity after eating clovers, buckwheat, rape, and Saint-John's-wort. Several conditions must be present for photosensitivity to develop: a light-colored goat, the kind of forage that causes the

Fig. 8-7. Don't tether a goat where it can reach any kind of poisonous plants such as the yew shown here.

ailment, and bright sunlight. Dark-colored goats are unaffected, and Saanens that eat the plants during cloudy weather are not harmed.

Protecting your herd against poisoning requires that you identify the poisonous plants that grow in your area and then be alert to any plants that are located where your goats can reach them. See that your goats are not exposed to poisonous plants after a long time in a dry lot or pasture that has been overgrazed.

PLANTS THAT POISON THE MILK

During the United States' early history many people developed a strange illness that old-time doctors learned to identify by a peculiar odor. The odor was caused by acetone on the breath. The illness was a great mystery for many years, but it was finally associated with drinking milk from cows that had the "trembles." It was then called milk sickness. One of the reasons for the difficulty in linking milk sickness with milk was that cows often produced toxic milk long before they showed the trembling symptom. The cow could show no symptoms for a considerable time until she had eaten a certain plant as a steady diet for several days. The substance that was toxic to humans was concentrated in the milk, so that people drinking the cow's milk were poisoned before the cow was affected. The name of the plant that caused many deaths before it was identified is *white snakeroot.* It sometimes goes by names such as richweed or Indian sanicle.

This plant is important to goat keepers only if the goats have access to areas where many wild plants grow. It grows near the edges of woods. Snakeroot multiplied rapidly on the land that the pioneers cleared. Many people died, and some regions were almost abandoned by settlers for a time.

Snakeroot is a branching herb about 2 to 4 feet tall. It has smooth oval leaves that are opposite on the stem. The leaves are broader and sharply toothed toward the base of the plant. The plant blooms from midsummer until frost, with tiny white flowers held in loosely branched clusters. The flowers occur at the ends of the branches. If you find a plant that answers this description in your pasture, get it out, or at least locate a friendly botanist who can identify it for you.

Chapter 9

Breeding

There is an old saying among livestock producers that the male is half the herd. With dairy animals a male can be even more important than that statement suggests. Even if you attach little importance to the selection of a buck to breed your doe, a questionable attitude if you value your herd's future, you still have to decide how you will obtain the services of a buck. The surest way of getting your does bred when you want is to own your own buck. This isn't a practical solution for many goat keepers.

OWNING A BUCK

Owning a buck makes sense if you live many miles from any other goat herd that includes a desirable buck. The lower the number of does you keep, the less desirable it is to keep a buck. Part of the reason is based on arithmetic. A buck must be fed, sheltered, and given the care of any other adult goat. If you have a single doe and add a buck, you are doubling your work and cost requirements. Good bucks are expensive to buy.

Your location with regard to other homes may help make a decision for you on keeping a buck. If you live in a crowded neighborhood, it's unlikely the neighbors would stage a block party to welcome a new buck. Even with descenting techniques, bucks develop an aroma during breeding season that appeals to the nostrils of does in heat but to few other creatures.

Mature bucks lapse into bad habits during breeding season that all but the most confirmed goat lovers find disgusting. The bucks urinate on their forelegs and face. These parts sometimes stay soaked with urine until the stench is noticeable hundreds of feet from the buck lot. If the buck is Saanen or other light-colored breed, the hair develops a yellow shade on the affected parts that sometimes darkens as dirt or other material clings to the damp hair. These habits are most offensive with bucks that are kept in a small pen during the breeding season. If you have limited space for your goat herd, you probably can't afford room for a buck that will be non-productive nearly all year. A buck should have enough space to be happy and sufficiently occupied to reduce his obnoxious qualities. If you have limited space, you probably will also have close neighbors.

Another disadvantage in keeping a buck on a small place is his odor can affect milk quality, unless the buck's pen is located a suitable distance from the does' quarters. The required distance varies according to the buck and season, but 50′ is a minimum, and 100′ to 200′ would be better.

If you have does of more than one breed and you want to breed each doe to a buck of her breed, then the disadvantage of owning your own buck is obvious. Even if they are all of the same breed, you must change bucks every two years or so to prevent inbreeding the buck with his daughters.

Bucks require taller and better fences than the typical doe. A good fence is essential for the buck pen, or any attempt at herd planning is doomed to failure. People have told me about a young doe that was bred prematurely because she got into the buck pen, or the buck got out while she was in heat. This seems to be commonplace even among experienced goat keepers with registered herds. A buck must be kept in a virtual cage of welded wire fencing if accidents are to be avoided. Fencing a suitable exercise area with welded wire for a buck is expensive.

These problems reduce the desirability of owning a buck. If you have the space and own more than a single doe, you may have reasons for buying a buck. One good reason is better disease control. You can have your buck tested and know that he isn't going to introduce brucellosis into your herd, although that is a remote possibility with any buck. You can also be complacent about other infectious diseases that can be picked up by visiting other goat farms with your does.

Perhaps the most appealing reason for buying a buck is you don't have to haul your does halfway across the county for breeding. Plus, you don't have to repeat the trip occasionally for a doe that didn't "settle."

Owning a buck is a decision that only you can make after considering your own situation and temperament. If you choose to keep a buck, prepare him for the breeding season by making sure he is wormed. Give him a ration with sufficient protein to permit good fertility. Most authorities recommend that a buck be fed the same kind of ration as milking does, at least through the breeding season. Before breeding, his belly should be trimmed so that long hair won't interfere with breeding.

BUYING A BUCK'S SERVICES

Buying a buck's services has many advantages. You can shop around for the ideal buck for each doe you own, and this means finding a buck with complementary characteristics and of the same breed. All the cost, care, and responsibility of the buck is met by someone else, while you keep only those animals that produce milk or meat. Remember this when the buck owner states the price for a service. It may seem expensive for the few minutes of time required, but it goes a short way toward paying the full year's cost of keeping a buck. Some stud farms offer a boarding arrangment in which your doe will be kept for a modest fee for a few days to catch the next heat period.

If you plan to use the services of a buck that belongs to someone else, decide on the exact buck well in advance of the heat season for your does. Visit the herds within easy driving distance to see which ones have healthy animals and use normal care in disease prevention. Some buck owners demand a health certificate from a veterinarian before breeding a doe. Don't be insulted if a buck owner is inquisitive about your does. The fact that he is choosy about the does that his bucks serve is a positive sign. Have the bucks and belly hair been trimmed? Is he in good condition generally? This doesn't mean that he should be fat. Bucks can look undernourished during the breeding season and be quite healthy.

ARTIFICIAL INSEMINATION

In some ways *artificial insemination* would seem to be the ideal way to obtain the services of a superior buck. Unfortunately it isn't available everywhere for dairy goats, and some artificial insemination that is available is poorly controlled. Artificial insemination has worked well for dairy cows. Semen from the best bulls in the country are available to almost any dairy herd. The practice is just beginning for goats. It may be the wave of the future after sire selection is more carefully done.

Some goat owners have tried to handle the insemination themselves by keeping frozen semen and breeding each doe as she comes in heat. There are so many problems that this seems highly impractical. The user must purchase a liquid nitrogen tank, which can cost hundreds of dollars. The tank must be periodically serviced with liquid nitrogen. Perhaps there is no one near you who provides such service.

Check the telephone directory to see if an artificial insemination service for goats is available in your area. Most of the artificial insemination people you ask will likely have no provision for keeping goat semen or inseminating does. If they do perform the service, try to learn something about the quality of the bucks they use. Be aware that artificial insemination doesn't always work the way it theoretically should; you may have to call the inseminator back for a second trial.

CHOOSING A BUCK

If you decide to keep a buck or rent one's services, you must decide on the specific buck for your doe. This decision is of more significance for future herd production than most people think. The buck generally contributes 50 percent to the next generation of does in a herd, but few people are aware that the buck can also affect the future production of the doe that he services. This may seem like superstition, but it is based on thorough research with dairy cows. One Norwegian study showed that cows mated to a beef bull produced about 13 pecent less milk than when mated with a quality dairy bull.

The reason is hormone production in the placenta during pregnancy. The placenta connects the dam and progeny during development of the fetus, but the placenta is genetically part of the fetus rather than of the dam. The buck has an influence on placental development and on the hormones produced in the placenta. The hormone it produces that favors milk production, placental lactogen, helps the dam's mammary system prepare for the next lactation period. The hormone also aids in udder development, so it probably influences future lactation periods. This suggests that young does should be bred only to the best bucks available if they are to reach top production figures.

Unless there is a special reason to do otherwise, try to find a quality buck of the same breed as your does, even if you have no plans for registration of offspring. One reason to avoid a buck of the same breed is that all the bucks of that breed are closely related to your doe. Another reason is to find a buck that is strong in the characteristics your does lack. If your does have udders that are too pendulous, look for a buck of any breed that has a reputation for producing daughters with good udder attachments. If milk production is your goal rather than registered offspring or show animals, crossbreeding can be a useful tool in herd improvement. Hybrid vigor can be used to contribute stronger, healthier does.

Should you use a registered buck or a good grade animal? The odds favor using a registered animal, because breeding records of registered stock will be more accurate and extensive. This information will help you choose a superior buck or at least one from superior stock. This isn't to say that all registered bucks are better than all grade bucks. Unfortunately some registered animals are too inbred for their own good or for the good of their offspring.

The ultimate quality of a buck can be obtained by inspecting his daughters. A superior buck will produce superior daughters, and all other characteristics are secondary to that one. He will also be more consistent in producing daughters of good dairy quality from many different does than other bucks. You should be able to check the quality and production records of many daughters to assess a buck's performance. A single daughter of high quality might suggest that a buck is a superior individual, but it proves nothing. When checking a buck's progeny, look for characteristics that show up time after time, indicating his influence. These characteristics will be good ones for a superior buck. The more proven daughters a buck has produced, the higher will be the stud fee.

You may have to use a buck without much more to assess than his appearance. Many goat owners just won't have enough does for you to inspect as many daughters as would be desirable. In such cases you can look for obvious virtues or faults. You probably would not choose a buck if he failed to show acceptable breed characteristics. A Saanen buck with a Roman nose would be a questionable choice. In comparing the conformation of various bucks, remember that a buck should have dairy character, but with a masculine appearance. His feet and legs should be sound and his body deep. The buck should look healthy. Disqualify any buck that looks as if he is diseased.

A superior buck will be large for his breed, because larger does usually produce more milk. His stance will be that of strength, with space between the hocks, suggesting that he will produce daughters with room for good udder development. A strong stance is evident when the animal seems to have firm underpinnings, with the feet firmly planted and the legs in a vertical position. It is common for the back legs to lean slightly forward. The head should have a masculine look, with full nostrils and a wide muzzle. The buck's body should be deep with the ribs fully sprung, showing large body capacity.

If the buck meets the general requirements, check that he shows only two rudimentary teats. One goat keeper bought a highly touted young buck that looked great. One day when he put his forefeet on his pen fence, though, she could see three small teats. A buck like this would probably produce more than his share of daughters with three teats, a highly undesirable characteristic. She called the breeder who sold the registered buck kid, and he immediately replaced the buck with another one of equal pedigree. Any reputable seller should do the same.

Check the reproductive organs of any buck that you intend to buy. The scrotum should be well developed with two testicles. Check the penal sheath for lumps or other symptoms of reproductive problems.

THE HEAT CYCLE IN DOES

Does can normally be bred once each year, and the time of breeding is seasonal. Most does will not come in heat under normal conditions until August or September. Unless the does are bred, they will continue to come in heat at approximately 21-day intervals until February or March. The heat period can vary from a few hours to a couple of days, but it usually lasts about 24 hours.

This seasonal breeding causes a problem for the goat keeper who wants a year-round supply of milk. Unless precautions are taken to avoid the normal timing of does, there will be a surplus of milk in the spring and a scarcity in the winter months. The doe will have access to the period of maximum plant growth soon after giving birth, so she can produce the most milk while her kids are growing rapidly and need it. You must try to "fool mother nature" to prevent this cyclical production rate.

One way to induce does to come in heat at odd times of the year is to change the ratio of light hours to dark hours in the winter. You can install fluorescent lights in the shed. Make sure they are exposed to 20 hours of light each day, starting in early January. Get a timer to control the lights when you are asleep or away from home. Set the timer each night to keep the lights on until four hours before sunrise. Continue this procedure for two months, then reduce the light period from 20 hours to 14 hours. This sudden change in light duration will often produce heat periods in the does, so you can breed some of them in the early spring for kidding in late summer or early fall. That will insure a winter milk supply. One drawback to this kind of tinkering with the seasons is you must expose the buck and the does if you want a reasonable assurance that it will work. Young, vigorous bucks will probably be able to service your does any time of the year without conditioning by light. The artificially induced heat periods are often very short, so you must be alert for them.

If you want to breed a doe in the summer, you can reverse the light treatment used in earlier months. Build a shed that is fairly "light-tight" and put the does in total darkness for about 17 hours per day. The shift to shorter daylight exposure should help bring them in heat during June or July. A doe bred in June will kid in November, producing her maximum milk flow in midwinter. The bucks to be used should be subjected to the same treatment, but young bucks may not need this kind of conditioning.

Beyond the obvious effort, cost, and time required to alter light exposure, artificially induced heat extends the work season of caring for young kids. Consider this before trying to make your does produce milk the year-round.

DETECTING HEAT PERIODS

A doe in heat will usually show signs of nervousness or at least restlessness. She may bleat more than usual and will often look over the fence as if seeking a buck. Tail wagging is a definite symptom. The sexual organs will usually be enlarged and more flushed than usual, with the possibility of a discharge. These symptoms vary considerably from doe to doe and are sometimes quite subtle. To verify a heat period, take the doe to a buck pen if that is convenient. The doe's symptoms will be greatly magnified at the sight and smell of the buck. If you have no buck in the neighborhood, keep a buck cloth during heat season. A buck cloth is nothing more than a rag that has been rubbed on the bucks scent glands and stored in a tight container. When you suspect a heat period, remove the rag and hold it near the doe's nostrils. If she responds with increased tail wagging or other symptoms, it is a further indication of heat.

Ovulation usually doesn't occur until near the end of the period when the doe shows symptoms of heat. This is considered a good time for breeding to obtain conception, but some goat keepers suggest that early heat period breeding is more likely to produce doe kids. Others suggest two breeding contacts with the buck, one near the beginning of the heat period and another 24 hours later. Your compliance with these rules will probably depend on whether you have a buck at hand or must travel a considerable distance to breed the doe.

A defined heat period is probably a sign of general health in your does. If the heat periods are weak and almost unnoticeable, it may be a sign of worms that can cause anemia. Worming the herd just before the breeding season reduces the problem. Some goat keepers use worming as a method of stimulating heat, perhaps because they have found that wormed goats show heat more readily than those that have been left with too many parasites. For those owners who follow a six-month worming interval, it makes sense to worm just before breeding and after kidding. Any doe that shows weak, barely perceptible symptoms of heat should be examined for worms.

AGE OF YOUNG DOES FOR BREEDING

In earlier times the conscientious goat owner usually tried to keep his doelings unbred until they were 18 months old. That practice is rarely followed now, because research shows that a doeling bred at 8 to 10 months will produce more milk in her lifetime than if the breeding were delayed a year. The doeling should weigh about 80 pounds before breeding, so a few slow developers may have to be held over an extra year. The possibility that the choice of a buck may affect future production suggests that you should obtain the services of the best buck available for a young doe.

GESTATION PERIOD

The average gestation period for goats is about 150 days. This is about five months, so you can get a rough estimate of the expected kidding date by counting off five months from the date of breeding. If the date of a doe's breeding is September 12, you can count off five months—October, November, December, January, and February. You can expect the doe to kid about February 12. This isn't exact because of the variation in the number of days in different months of the year.

To make a more precise estimate of the kidding date use Tables 9-1 and 9-2. Refer to Table 9-1 and get the key letter and month of kidding. For the example discussed earlier the key letter is C for September, and the month of kidding will be February. Find the day of breeding under C in Table 9-2 and read the day of kidding from the right-hand column. Placing a ruler across the chart will aid in finding the date accurately. In this case it is February 8. There is an obvious advantage in knowing the exact date if your doe has a

Table 9-1. Gestation Table for Month. To Find the Date of Kidding, Use This Chart with Table 9-2 as Directed.

Month of Service		Month of Kidding
Month	Key	Month
January	B	June
February	A	July
March	C	August
April	C	September
May	C	October
June	C	November
July	C	December
August	C	January
September	C	February
October	B	March
November	B	April
December	B	May

gestation period of average length. By roughly figuring the kidding date to be February 12, you might fail to be ready for the actual event, although the doe would probably show symptoms to alert you.

In a few cases the month of kidding must be shifted forward one month from that obtained from the chart. Perhaps the doe had been bred on September 1. Going by Table 9-1, you would expect the month of kidding to be February. When you look at Table 9-2, notice that a doe bred on September 31 will be expected to kid on February 27. A doe bred on September 1 would seem to kid on February 28. The actual date of kidding for a doe bred on September 1 would be January 28 and not February 28. The rule in using the charts is to shift the kidding a month back for any number in column A that occurs below 28, or any number in columns B or C that occurs below 31. This is necessary because the date 150 days after the first few days of each month is likely to occur during the last days of the fourth month, instead of the fifth month, after the month of breeding.

When the doe is bred, make a record of the date, the buck that was used, and the expected date of kidding. Table 9-3 is such a chart.

CARE OF THE PREGNANT DOE

Most does will give less milk shortly after being bred. Even those does that continue to milk well into pregnancy should be turned dry about two months before the kidding date. If a doe is a

Table 9-2. Gestation Table for Day of the Month. To Use, First Find the Month of Service in Table 9-1. From it You Can Get the Correct Column Key for the Month of Service and the Month of Kidding. Then Locate the Day of Service and Move Horizontally to the Date of Kidding.

Date of Service			Date of Kidding
A	B	C	
2	3	5	1
3	4	6	2
4	5	7	3
5	6	8	4
6	7	9	5
7	8	10	6
8	9	11	7
9	10	12	8
10	11	13	9
11	12	14	10
12	13	15	11
13	14	16	12
14	15	17	13
15	16	18	14
16	17	19	15
17	18	20	16
18	19	21	17
19	20	22	18
20	21	23	19
21	22	24	20
22	23	25	21
23	24	26	22
24	25	27	23
25	26	28	24
26	27	29	25
27	28	30	26
28	29	31	27
1	30	1	28
	31	2	29
	1	3	30
	2	4	31

Table 9-3. Suggested Breeding Chart.

Name of Doe	Buck	Date of Service	Date of Kidding	Comments
Maisie	Oak Leaf George	9/28/82	2/24/83	Keep doe kids
Cleo	Oak Leaf George	11/17/82	4/15/83	For sale, bred

persistent milker, eliminate the concentrates in her diet. If she is producing less than ½ gallon of milk per day, you can stop milking. If she is producing more than ½ gallon, you should feed her a dry grass hay and limit water intake. Stop milking the doe for a week, then milk her out one last time. The milk production should ideally be at a minimum, a few ounces a day, when you stop milking her.

To keep the pregnant doe in good health, see that she gets adequate forage and enough exercise. Avoid too much concentrate in early pregnancy to prevent her from getting fat. When fat, a doe has poorer appetite than when in normal flesh. A fat doe is more likely to have ketosis at the time of kidding. Keep the doe's appetite strong by moderate feeding and by providing adequate exercise. This is one advantage of late kidding. A doe that will kid in midwinter may not get enough exercise in northern climates where heavy snow confines the goats to a shed for long periods. Exercise the does by walking them back and forth in the shed.

Add concentrates to the doe's diet as the time of kidding nears. A feed with molasses in it is a good way to supply enough energy in the ration to avoid ketosis. Increase the feeding ration slightly in the final weeks before kidding. Rapid fetus growth during this period requires more concentrate to supply enough protein. Reduce the grain ration in the last days of pregnancy. Add bran to the feed during this period to provide the doe with a mild laxative. During the last week or so before the due date, clip the hair around the udder, tail, and hindquarters.

The udder will enlarge during the last weeks of pregnancy. When the udder fills so it is tight and shiny, the day of kidding is at hand. The doe's sides will show deep hollows. She may have a slight discharge and will often paw the ground. The specific symptoms vary from animal to animal, but the general attitude is one of restlessness.

KIDDING

To prepare for kidding, provide the doe with a private space by placing a folding gate or other barrier across a small area in the shed. The floor of the kidding space should be covered with fresh, clean bedding material. Remove any objects that can interfere with kidding such as boxes, feeders, or buckets. If the temperature in the shed is below freezing, place a heat lamp above the kidding space if one is available. If you have no heat lamp, monitor the kidding process more closely so that you can help the doe dry the kids as soon as they are born. Don't crowd the doe unnecessarily. She will likely give birth normally with no help from you.

A doe that is in good condition rarely has trouble kidding. One of the reasons for providing adequate exercise during pregnancy is to keep the doe in proper condition for this moment. If a doe strains hard for an hour without delivering, investigate the problem. Call a vet or someone with experience in helping deliveries.

Various problems can cause the doe to have trouble delivering alone. If she is a young doe with a single kid, the kid can be too large for the doe to deliver it. A veterinarian will have to do a cesarean. Call the veterinarian before the doe becomes exhausted and in poor condition.

A kid's head may be twisted back, blocking exit. This requires a hand in the uterus to straighten the kid for normal birth and is best done by someone with experience. If no one is available, wash your hand and arm carefully with soapy water, put on a sterile rubber or plastic glove, and feel into the birth canal with your fingers to see if you can locate the problem. Kids are normally born head first, with the head between the forelegs, or rear feet first. If everything seems right, you can aid the doe by pulling with a steady tug backward and down. Kids born rear end first may need help, because the umbilical cord may be broken while the kid is still in the birth canal. Its nostrils must be cleared soon so it can breathe. As soon as the kid is out, pick it up by the rear legs. Slap it sharply on the sides to help clear the breathing passages and establish normal breathing. This is not necessary with kids that are born head first.

Dry the newborn kid with a clean towel. The doe will lick the kid dry (Fig. 9-1), and that is usually adequate in warm weather. Dip the kid's navel cord in a jar of tincture of iodine. The cord should be submerged completely for proper treatment. Wash the doe's udder with a warm soapy solution, then rinse off the soap and dry the udder. Provide the doe with about 1 quart of warm water in which ¼ to ½ cup of molasses has been dissolved. Don't let her drink cold water. She will have lost body heat in the kidding process. Her drinking water should be at least as warm as normal body temperature—about 100 degrees Fahrenheit or higher.

After the doe has finished kidding, watch for the afterbirth. If she doesn't pass it within several hours, call the vet. If you go away and come back hours later to find no afterbirth, you can't know for sure that she hasn't passed it. Some does eat the afterbirth.

RAISING KIDS

If you plan to immediately separate the kids from the doe, milk

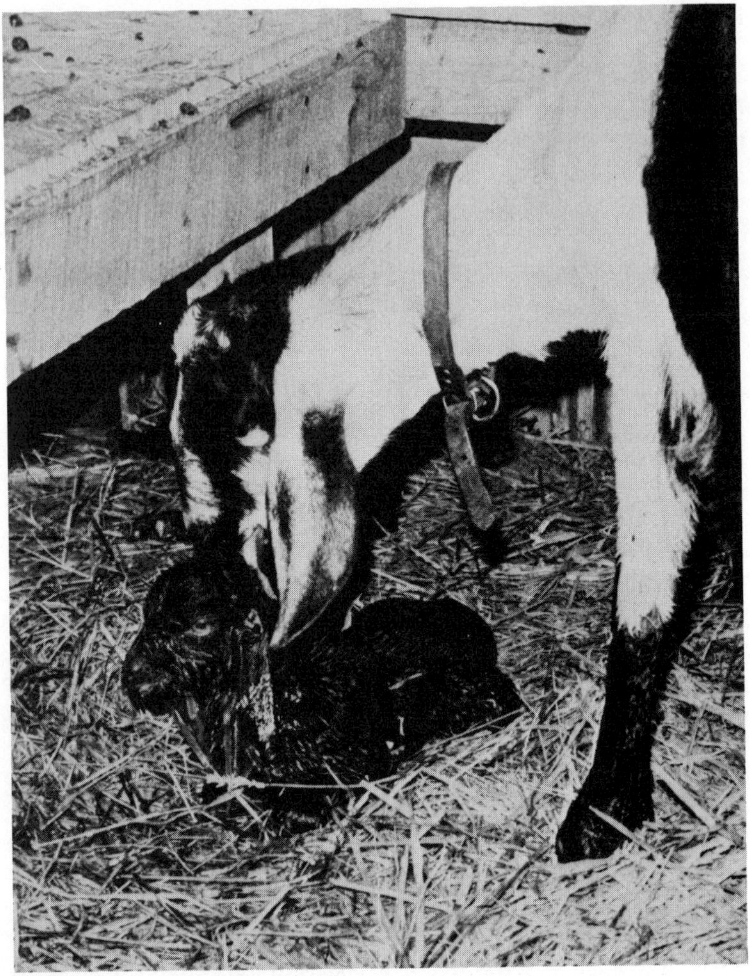

Fig. 9-1. A doe licks a newborn kid.

out the udder enough to get at least 1 pint of milk. Put the milk into sterilized nursing bottles and feed the kids. Each kid must eat a good supply of colostrum milk. Colostrum is high in protein, lactose, and solids. It contains *gamma globulin* from the doe and gives the kids resistance to any disease to which the doe is resistant. Kids with no colostrum usually die within three days from white scours or some other infectious disease. A normal kid should have about ½ cup of colostrum at the first feeding. The sooner the kid gets the colostrum after birth, the better. Young kids need feeding at frequent intervals—at least three times a day. Feeding kids several times the first day is a good way to get them to absorb extra colostrum, which protects the kid until it is old enough to build up resistance against disease on its own.

What do you do when a doe dies while giving birth or immediately afterward? You should have some extra frozen colostrum in your freezer. Unless the kid gets enough colostrum the first few days, it will probably contract disease.

If you need the goat milk for family use, you can shift the kids to milk replacer, available from feed stores, after about a week. Make the shift gradually. Do not use milk replacer before the kids are a week old. You won't want to use the milk from a freshened doe until the colostrum is milked out completely, and the milk is normal.

Some goat keepers prefer to leave the kids with the dams for at least a short period after kidding. They say there are fewer problems with scours or other infections when the kid gets the milk directly from the udder rather than through various utensils that can contain bacteria. These goat keepers note that it's easy for the milk to cool from goat body temperature, and cold milk can cause scours. Proponents of early separation say the doe will suffer no emotional shock if the kids are taken away the first day, but later removal can cause her to drop in production. They also mention that kids raised by bottle feeding are tamer and easier to handle as adults. Some owners allow the kids to graze with the doe during the day. The kids are separated from the doe at night so that the owners can get milk for their families in the morning. The doe's udder should be milked out once a day anyway. People who favor separation suggest that large kids can injure the udders of high quality does by tugging at them all day and by frequently butting them. Goat owners who permit kids to run with the dam report that they have few problems.

One goat owner that I know uses a compromise plan. She separates the kids from the does at an early age, but allows the kids to nurse the does twice a day. She turns the kids into the milking room while the does are eating, after obtaining enough milk for family use (Fig. 9-2). She says this gives her the best of both worlds. The kids do much of the milking labor and yet they can't follow the doe around all day. She doesn't have to struggle with keeping the milk at 103 degrees Fahrenheit for young kids.

If you separate the kids, start them early on a concentrate such as calf starter. Kids will usually learn to eat green plant material before they will take hay, so a kid pasture is desirable if you have the space. Any kind of green pasture will do except one that contains

Fig. 9-2. A low-labor method of milking and feeding kids.

poisonous plants. Roughage that the kids like will help develop rumen capacity, and that prepares the does for better production when they mature.

KIDS TO MARK FOR MEAT

You will probably want to use all your male kids for meat or sell them to someone else for that purpose. The world needs no more bucks of questionable quality, so castrate the males when they are a week old. Examine your female kids carefully as soon as you can estimate their degree of dairy character. Obviously you will want to butcher or sell for butchering any females that have serious faults such as hermaphroditism. Each doeling that you keep must be of good enough quality for your herd. You will be feeding each one for at least a year before getting any production from her. It doesn't pay to feed an inferior animal that long except for the meat that she may produce, either by having kids or being butchered.

Here is one point where goat keepers are likely to compromise. All kids are cute. You have to examine each one as to how well it meets the requirements of your family for milk and meat. Give every goat a good life by proper care while you own it, but don't feel you have to give it a pension. If you plan to sell kids such as surplus wethers to other people for meat, sell them when they reach the stage where they can live on milk replacer.

PROTECTING KIDS FROM PREDATORS

In some areas kids must be protected from dogs or coyotes. Dogs are usually the greater hazard in towns. Several dogs running in a pack often develop a taste for blood. Kids are especially vulnerable. Large dogs will attack mature does and sometimes even bucks.

Most cities have laws about dogs running loose, but many small towns don't enforce these laws. Recently I was lying in bed, reading, when my wife asked, "Why are all the dogs in town barking?" We listened for a minute, and over the sound of the dogs was a more plaintive sound. I went to the back door to check our goats, but all was quiet in that direction. When I opened the front door, I could hear a goat crying at a neighbor's place about a block away. I could see lights at the house, so I went to the telephone and called my neighbor John. When I asked what was wrong at his place John said, "Nothing. Just watching television."

I told him to check his goat pen. He found that three or four dogs were pulling Peace Valley, his best young doe, off an exercise stand in the goat lot. The other goats were huddled in the barn. He grabbed a stick and beat the dogs off Peace Valley. Peace was hurt badly, with a gaping hole in her side. Peace had to be nursed for months and was not able to be bred during the season, losing a year of production. If John had gotten there a few minutes later, she would have been dead.

Fortunately he was able to recognize the lead dog and notified the owner the next day. The owner of that dog was cooperative, but many dog owners will refuse to believe that their dogs would ever harm anything. Although that owner disposed of his dog, the carnage continued in our town until one man lost about six goats, another family lost a flock of chickens and some rabbits, and a nearby farmer lost 28 hogs. At that point someone took the law into hand. Residents suddenly noticed fewer dogs around town. A boy out for a country hike found a ravine into which a pile of dogs had been dropped. The reign of terror ended, and goat keepers could sleep again at night. Such outbreaks usually occur when the population of stray dogs gets out of hand or when land developers bulldoze large acreages, reducing the animal population and sending coyotes into populated areas for their prey.

A good fence is your first line of defense. If the fence is tight and high enough, most dogs will hesitate to cross over it. A few large dogs will jump fences that are as high as 4′. If you have such dogs in your neighborhood, put your goats into a closed shed at night.

Dogs or coyotes will occasionally dig under fences. To prevent this digging, you can lay a woven wire fence along the ground. Attach the woven wire fence to the bottom of the upright fence and cover it with soil. This adds tremendously to the cost of fencing, so you would probably want to use such an extreme measure for small pens only.

A guard dog will defend your goats from predators (Fig.9-3). These dogs are expensive to buy, but they are on the job day and night. Guard dogs are the most practical solution if you have a large

Fig. 9-3. A Komondor on guard.

feel you have to give it a pension. If you plan to sell kids such as surplus wethers to other people for meat, sell them when they reach the stage where they can live on milk replacer.

PROTECTING KIDS FROM PREDATORS

In some areas kids must be protected from dogs or coyotes. Dogs are usually the greater hazard in towns. Several dogs running in a pack often develop a taste for blood. Kids are especially vulnerable. Large dogs will attack mature does and sometimes even bucks.

Most cities have laws about dogs running loose, but many small towns don't enforce these laws. Recently I was lying in bed, reading, when my wife asked, "Why are all the dogs in town barking?" We listened for a minute, and over the sound of the dogs was a more plaintive sound. I went to the back door to check our goats, but all was quiet in that direction. When I opened the front door, I could hear a goat crying at a neighbor's place about a block away. I could see lights at the house, so I went to the telephone and called my neighbor John. When I asked what was wrong at his place John said, "Nothing. Just watching television."

I told him to check his goat pen. He found that three or four dogs were pulling Peace Valley, his best young doe, off an exercise stand in the goat lot. The other goats were huddled in the barn. He grabbed a stick and beat the dogs off Peace Valley. Peace was hurt badly, with a gaping hole in her side. Peace had to be nursed for months and was not able to be bred during the season, losing a year of production. If John had gotten there a few minutes later, she would have been dead.

Fortunately he was able to recognize the lead dog and notified the owner the next day. The owner of that dog was cooperative, but many dog owners will refuse to believe that their dogs would ever harm anything. Although that owner disposed of his dog, the carnage continued in our town until one man lost about six goats, another family lost a flock of chickens and some rabbits, and a nearby farmer lost 28 hogs. At that point someone took the law into hand. Residents suddenly noticed fewer dogs around town. A boy out for a country hike found a ravine into which a pile of dogs had been dropped. The reign of terror ended, and goat keepers could sleep again at night. Such outbreaks usually occur when the population of stray dogs gets out of hand or when land developers bulldoze large acreages, reducing the animal population and sending coyotes into populated areas for their prey.

A good fence is your first line of defense. If the fence is tight and high enough, most dogs will hesitate to cross over it. A few large dogs will jump fences that are as high as 4′. If you have such dogs in your neighborhood, put your goats into a closed shed at night.

Dogs or coyotes will occasionally dig under fences. To prevent this digging, you can lay a woven wire fence along the ground. Attach the woven wire fence to the bottom of the upright fence and cover it with soil. This adds tremendously to the cost of fencing, so you would probably want to use such an extreme measure for small pens only.

A guard dog will defend your goats from predators (Fig.9-3). These dogs are expensive to buy, but they are on the job day and night. Guard dogs are the most practical solution if you have a large

Fig. 9-3. A Komondor on guard.

goat herd in an area menaced by coyotes. Jim Wendland, who owns a herd of more than 100 does in Arkansas, found that guard dogs were the only way he could protect his goats. He tried every remedy that was suggested. He even hung his dirty T-shirts on fences in hope that the human odor would deter the predators, but the coyotes kept killing his goats. Finally he bought some Komondor pups at $450 each. The *Komondor* is a guard dog brought to this country from Hungary. The Komondors spend day and night with the goats (Fig.9-3). Jim has lost no more goats to coyotes. The Komondors seem like jolly giants, but they are all business when guarding goats.

If you are bothered by any kind of predator, learn what recourse you have by law before doing anything. Even if the law permits you to shoot dogs that are killing your goats, you may have trouble with the dog owners. This is one of those problems that you have to solve according to the peculiarities of your own situation.

Chapter 10

Produce: Milk and Meat

The payoff in keeping goats is the milk and meat that they can produce for your table. Although most people who have goats want the milk for family use, it can be useful on a small farm for feeding other farm animals such as chickens or pigs. Many goat owners buy three-day-old calves from dairymen and raise the calves on goat milk. This can be a real profit making venture for goat keepers. Almost everyone has a few pets, and surplus goat milk helps reduce food bills. Regardless of your uses for milk, the quality of the milk you produce should be as high as possible.

FACTORS AFFECTING MILK QUALITY

A program for milk quality improvement starts with the milking room. It should be kept as clean as possible and be separate from the space used by goats at any time other than milking hours. If you have no shed space for an adequate milking room, do the milking chores outdoors, away from manure or other materials that produce odors which can be absorbed by the milk. Milk is like a sponge to undesirable odors, so ventilation and cleanliness are essential.

To reduce the chore of cleaning does before milking, make sure the barnyard or pens are well drained. Does that have to wade through mud or manure to get to the milking room will carry in filth that can affect milk quality. The time required to clean them will be increased tremendously.

If you keep pigs or chickens on your place, they should be

housed at least 50′ from your milking room and preferably farther. The odors given off by these animals will vary according to how you keep them. Pigs are not dirty animals by choice, although they love to loll in pools of water that often are reduced to mud baths. When pigs are kept in a tiny space, they invariably produce odors. Chickenhouses are notorious for odors, especially where the chickens are crowded.

Prepare for milking by wiping the doe's udder with a clean, damp cloth. If you want to be especially careful, you can use a special disinfectant in warm water. Wiping alone removes the dust and fine bits of straw or manure that can cling to the udder or teats. This material will add greatly to the milk's bacterial content if it falls into the milking pail. Make sure your own hands are clean and dry before you begin. Use a scalded or disinfected milk pail with preferably no seams in the metal. Stainless steel pails are expensive, but they will last a lifetime. Don't use plastic containers for milk. Plastic is almost impossible to clean adequately, and some plastics impart a flavor of their own to the milk. The ideal milk pail has a restricted opening at the top to prevent contamination by falling hair or bits of manure or hay.

Remove the milk from the shed area as soon as possible. Some goat keepers prefer to milk into a glass jar that has a tight-fitting lid. They milk the goat, then screw on the lid to protect the milk from odors until it can be refrigerated.

Sediment of the kind often caused by careless milking practices can be removed from the milk by straining, but you can't remove the bacteria introduced by the sediment that way. The emphasis must be on preventing contamination from reaching the milk and not on removing it. You can kill bacteria and destroy enzymes by pasteurizing milk. Strain and pasteurize the milk immediately after milking. Strain the milk through sterile cotton filter disks available from farm supply stores. Some people strain the milk through reusable cloths, which are then washed. This does remove foreign matter, but it is less desirable than the obviously superior method of using a sterile pad.

If you use a fine metal strainer for milk, it should be a stainless steel strainer. Copper is especially bad in producing changes in milk that cause a tallowy flavor, so avoid any contact with copper or copper alloys. Copper increases the rate that fats in the milk are oxidized. Once the process of oxidation begins, it feeds on itself. Other metals can have a similar effect.

You can easily lose quality in the milk your does produce by careless feeding. The most obvious way to compromise quality is to allow strong tasting weeds to grow in your goat pasture. One of the worst weeds is bitterweed or *Helenium tenuifolium.* This weed is insidious because you can't smell the bitter flavor it induces in the milk. No amount of boiling the milk will remove it, because the flavor material isn't volatile. Ragweed is notorious for producing strong flavored milk. Goats seem to like the taste of giant ragweed—sometimes called horseweed or *Ambrosia trifida.* Other weeds such as wild onions can affect milk quality, and you can assume that any weed which has a strong or noticeable odor when crushed will flavor milk.

Some feed concentrates can cause off-flavors. This is one reason to choose dairy rations for goats and to avoid feeds formulated for other purposes. Some goat keepers like to milk their goats before giving them feed on the theory that the feed can flavor the milk almost immediately, but this is a questionable procedure for goats. Most goats will be made unhappy by this practice. If you hold back feed, they'll hold back milk.

Milk can develop a rancid flavor when kept for several days. This is often caused by the enzyme *lipase.* It induces hydrolysis of the milk fat, producing butyric acid and other volatile acids. You will find this activity more noticeable as the doe approaches the end of her lactation period. Pasteurization will deactivate the enzyme and prevent this deterioration.

PASTEURIZATION

Any milk used for babies or small children should be pasteurized, because they are more susceptible to bacterial infections. Some people claim raw milk has various qualities that make it more healthful, but research has rarely shown any difference between the value of raw milk and pasteurized milk.

The amount of time that milk must be held at the pasteurization temperature depends on the temperature reached (Table 10-1). Real Fresh, Inc., of Visalia, California, has begun to market milk subjected to a process they call UHT—ultrahigh temperature—that permits you to store the milk in unopened cartons on the shelf for six months with no refrigeration. To produce such resistance to spoilage, Real Fresh subjects the milk to a temperature of 300 degrees Fahrenheit for three seconds. This process probably requires special equipment for rapid heating and cooling.

You can buy a home pasteurizer or heat the milk on top of your range. For pasteurization either heat the milk to 142 degrees

Table 10-1. Time Required for Pasteurization at Various Temperatures. For Safety, Add 1 Degree to These Readings.

Temp. (Fahrenheit)	Time
160°	15 seconds
155°	60 seconds
150°	3-1/2 minutes
145°	13 minutes
142°	30 minutes
140°	55 minutes

Fahrenheit for 30 minutes or to 160 degrees Fahrenheit for 15 seconds. In these days of high energy costs the second method seems more practical, but you must have an accurate thermometer to make sure you are getting the milk up to the critical temperature. One way to check yourself is to use a cheap candy thermometer to monitor the pasteurizing process and check it occasionally with a more accurate one. If you have a high school science student in your family, let him or her take your thermometer to the school lab and check it against one of the lab thermometers. You can buy a Kodak color thermometer, designed for use with temperature-critical color film processing, and keep it in a safe place. Use it only to check the accuracy of the thermometer you use to test milk temperatures.

While heating the milk for pasteurization, stir it constantly so that hot spots don't develop. If the temperature of any part of the milk goes above the required temperature for a time longer than that needed for pasteurization, the milk may develop a cooked flavor. Pasteurizing the milk will likely cause a slight change in flavor. The flavor won't change with storage time as it would if you did not pasteurize the milk.

As soon as the milk is held at the prescribed temperature for the time required for pasteurization, chill it as quickly as possible. This means a cold water bath around the milk container, with frequent agitation of the milk and the water bath. Change the bath as necessary to keep the temperature differential between the milk and the bath great enough to transfer the heat from the milk into the water. Milk should be stored at about 40 degrees Fahrenheit. An ice water bath is even better, because the speed of cooling determines how much flavor change pasteurization will make. To reduce the chances of a cooked flavor, instant chilling is best.

MILKING GOATS

Most goats are easier to milk than cows. The process of cleaning the udder helps to set up conditions for milk flow. Wipe the udder with a clean damp cloth or use an udder disinfectant. Wipe it dry with a clean paper towel. Use a different cloth and a new towel for each goat. By doing this and further massaging the udder, the doe will involuntarily let down her milk.

To milk, grasp a teat in each hand and squeeze, starting with pressure from your thumb and forefinger at the upper end of the teat. Add pressure in sequence from your other fingers down to your little finger. Note that you don't have to move your hands on the teats; simply follow the flow of milk down the teat with a rippling action in your fingers and palm. I know of only one person who had trouble with this kind of action. She says that when she squeezed, the milk went back up into the udder and the doe stood on her hind legs.

Most people set up a rhythm of left hand, right hand, left hand, right hand. This has a subtle advantage. If the doe decides to protest your milking by kicking or raising a foot, the flow from one teat to another keeps her off balance. To further convince her that you mean business, you can press your head into her side. Any restraint keeps her from trying to get away. The doe should be restrained by a stanchion or a collar and chain, so she can never break and run. Once the doe does this, she may try to escape again and again. This is particularly true of young does who haven't been milked.

When you have emptied the udder by the normal squeezing process, place your thumb and forefinger across each teat and gently strip out the remaining milk in the teat. Some goat keepers prefer to do no stripping. They believe it may be harmful to the teat, but that seems unlikely unless the stripping is too vigorous. The advantage is that it removes any milk that would otherwise be retained in the teat, and retained milk is undesirable.

Try to schedule your milking times so that they come at regular intervals. You should ideally milk at the same hour every morning and evening. Unless you never leave home, there will surely be exceptions to this rule, but remember that any change in the schedule can upset a doe and reduce production. Goats are more habit prone than almost any other domestic animal.

In addition to milking at the same hour each day, try to follow the same milking procedure. If doe A is usually milked before doe B, don't suddenly change the routine to milking doe B first, unless there is a good reason to do so such as doe A getting mastitis. Any change will cause problems and likely reduce production in both does. The normal procedure is to milk the doe that is at the top of the pecking order first.

RECORD KEEPING

There are many reasons for keeping records on your dairy goats. By controlling the amount of feed to each goat and the weight of milk from each, you can use a challenge feeding program to get the highest possible production from each individual. There is no way to do this accurately except with complete records. One typical chart is shown in Table 10-2, but the exact form you use isn't important. You will need a set of scales in the milk room to weigh the output of each doe. Volume measurements aren't accurate enough unless you have a tall graduated cylinder, and that makes an extra container to clean.

Complete records may bring some surprises when you total the results of a year's production. A doe that appears to be producing less milk than another may have a higher yearly production because of greater lactation persistence. The records will show which does to keep and which ones to eliminate. Any doe that has a lactation of six months or less should be culled.

USING GOAT MILK

Goat milk can be substituted in any recipe that calls for whole milk. If you want to make butter, it will be necessary to separate the cream, and that is usually difficult with goat milk. One way to facilitate the separation is to add a small amount of cow milk. This adds the agglutinating protein of cow milk that makes the fat globules coagulate and rise. You don't have to use whole milk from cows to get the protein that makes cream rise; skim milk will do.

HOMEMADE GOAT CHEESE

Goat cheese, long a standard product in Europe and other parts of the world, can be made at home. Go to a health foods store and get some cheese rennet tablets. With Hansen's cheese rennet and the instructions given here, you can easily make Colby cheese and age it to your taste. You can make other kinds of cheese, but Colby is a good one for the beginning cheese maker.

Use nothing but pasteurized, high quality milk for cheese making. Unwanted organisms can produce unpredictable flavors as

Table 10-2. Suggested Chart for Feeding and Milk Record. X—Don't Save Milk.

Name		Maisie			Cleo		
Date		Feed	Milk Wt.	Status	Feed	Milk Wt.	Status
4/20	7:00 A.M.	1-½ lb. chow	5 lb.	O.K.	2 qts. oats	3-½ lb.	X
	6:30 P.M.	1-½ lb. chow	4 lb.	O.K.	2 qts. oats	3-¼ lb.	X
4/21							
4/22							

the cheese ages. The recipe that follows is based on using 2 gallons of milk, but you can vary the amount to suit your milk supply and your needs.

Collect 1 gallon of fresh milk in the evening and pasteurize it. Allow it to cool to about 50-60 degrees Fahrenheit and let it remain at that temperature overnight. Taste it the next morning to make sure there are no off-flavors or sourness. Add another gallon of freshly pasteurized morning milk, mixing the milk in an enameled or tinned pan. Adjust the temperature of the mixed milk to 86 degrees Fahrenheit. Add 1 cup of buttermilk, keeping the temperature to 86 degrees Fahrenheit. Divide a Hansen's cheese rennet tablet into four pieces and crush one-fourth of the tablet, so it will dissolve in a glass of cold water. Place the 2 gallons of milk in a water bath warmed to about 88-90 degrees Fahrenheit (Fig. 10-1). The water bath should be in a draft-free place.

Add the rennet solution to the milk, stirring gently but thoroughly for 30 seconds to a minute after adding the rennet. As soon as you finish this initial stirring, leave the milk undisturbed for at least 30 minutes or more for the curd to form. Test the curd by sticking your finger into it and pulling up. If the curd breaks clean around your finger and leaves a sharp edge, it is ready to cut. Cutting the curd helps to determine the hardness of the final cheese. If you want a hard cheese, cut the curd into small pieces; for a softer cheese, cut the curd into larger pieces. Hansen's Laboratory recommends that you cut the curd by holding a long knife in a

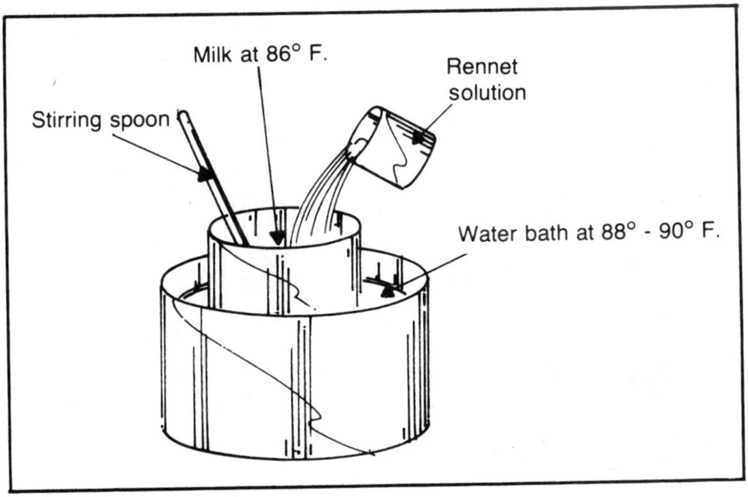

Fig. 10-1. How to use a water bath to control curd temperature.

vertical position and making parallel cuts across the curd, from top to bottom, ⅜″ apart. Then make vertical parallel cuts at right angles to the first cuts. Holding the knife at an angle, make diagonal cuts at ½″ to 1″ intervals. Finally, turn the container 180 degrees and hold the knife at the same angle as before to make another set of angled cuts (Fig. 10-2).

Heat the water bath to slowly bring the temperature of the curd up to 102 degrees Fahrenheit. The purpose here is to heat the curd until it holds its shape but falls apart readily when held in your hand without squeezing. Heat for an hour at 102 degrees Fahrenheit.

Bail out the whey and add enough cold water to reduce the temperature to 80 degrees Fahrenheit. Stir for about 15 minutes. The cooler you make the curd at this point, the drier the cheese will be. Place the curd in a colander and let it drain for about 15 minutes. Add 3 to 5 tablespoons of salt and mix thoroughly. Allow the curd to

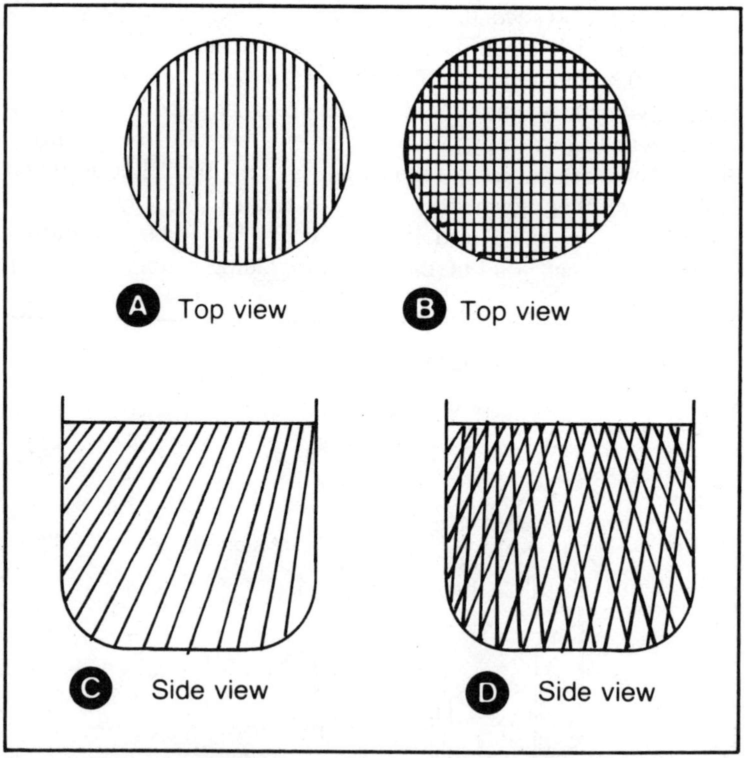

Fig. 10-2. Sequence followed in cutting the curd. (A) Vertical cuts, ⅜″ apart. (B) Vertical cuts at right angles to first cuts. (C) Angled cuts, ½″ to 1″ apart. (D) Angled cuts in opposite direction to C.

drain in a colander again for about 20 minutes. Test the cheese by biting it. If the cheese squeaks when you bite, it is ready for the press.

Place the cheese in a press, add 1 gallon of water to the bucket for weight, and leave it for 15 minutes. Remove it from the press and dress the cheese. To do this, wrap it with a strip of folded cheesecloth. Place cheesecloth above and below the cheese in the press. Place one follower below the cheese and enough followers above the cheese to prevent the board on the press from bottoming. Add another gallon of water to the bucket on the press. It may be advisable to add the water in ½-gallon lots, waiting a few minutes after each addition for the press to work. After 10 minutes, add another ½ gallon of water, making 2-½ gallons in the bucket.

If you make your press according to the instructions that follow, you will be applying a force of about 20 pounds at the end of the lever which, with a 2:1 mechanical advantage as shown, will equal 40 pounds on the cheese. If the press is moved farther out from the wall, add 3 or 4 gallons of water to the bucket.

After four hours, remove the cheese from the press. Coat it with a layer of Crisco shortening. Crisco shortening has a mold inhibitor in it that will offer some protection for the cheese while it is aging. Most instructions call for dipping the cheese in hot paraffin, but that is a more complicated procedure. Age the cheese in your refrigerator for three or four weeks before eating it. Colby cheese has a mild flavor and works very well in pizza recipes.

MAKING A CHEESE PRESS

The usual plans for homemade cheese presses require a baseboard with dowels projecting upward to hold an upper board, which is weighted with bricks to make it slide down the dowels and press the cheese. Some presses are a bit unstable. All are tricky to make because the holes in the top board must fit the dowels fairly well, but they must remain loose enough so the board will slide down the dowels. If one side of the board shifts too much, the dowels can bind and stop the press. The press described here is easy to make and works well. It was designed by Bill Keith, a dairy farmer and physics teacher, who makes cheese for his family on rainy days when he can't work outdoors.

You will need an empty vegetable shortening can, a 30″ board, several board disk followers cut to fit loosely inside the can, a large screw hook, and a bucket to make the press (Fig. 10-3). Fasten a cleat to your kitchen wall, or where you make cheese, above a

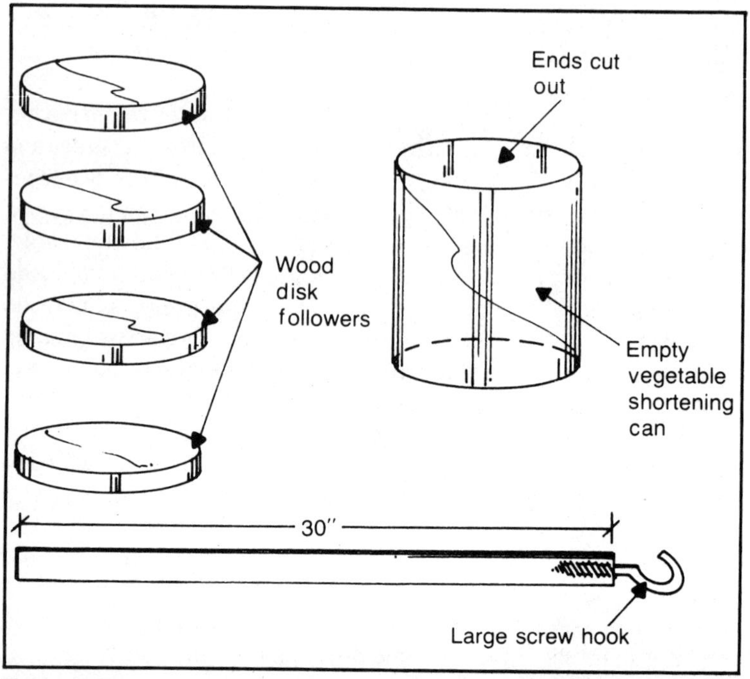

Fig. 10-3. Parts for cheese press. Diameter of wooden disk followers is made to allow disks to slide inside an empty vegetable shortening can.

counter or table. Screw the large hook into one end of the 30″ board. Remove the bottom of the can with a can opener, so it has no top or bottom. When you are ready to press the cheese, place one follower in the bottom of the open can, the cheese above it, and enough followers above the cheese to prevent the board from bottoming. The cheese will be wrapped in cheesecloth on all sides for the final pressing. Put the press in a pan to catch the liquid that is forced out of the cheese. Place the board over the followers with one end under the wall cleat. Hang a bucket on the hook at the other end (Fig. 10-4) and add water as required to press the cheese. Add the water in increments—not all at once. Add some water, wait 5 or 10 minutes, then add more. By using water as the weight, the amount of force applied to the cheese is continuously variable.

COTTAGE CHEESE

If you have more than enough milk for other uses, you can make cottage cheese. The same precautions about milk quality that were mentioned for making Colby cheese apply to cottage cheese. You can't make good cottage cheese from milk that has been stored for days.

Cottage cheese is made from skimmed milk, so you will have to add some skimmed cow milk to your goat milk to help separate the cream. Save the cream for making butter, but don't store it for too long before churning.

You can choose between two methods of making cottage cheese that will produce two kinds of product—small curd or large curd cheese. Many farm families used to make small curd cheese in the days when small farmers were self-sufficient. It has a more acid taste than the cheese you buy at the supermarket. You can use fresh buttermilk or a commercial lactic culture to make it. You will need some commercial rennet to make large curd cheese.

Because the process of making cottage cheese is quite similar to that of producing Colby cheese, the same kind of equipment is needed. Use a water bath around the milk container to maintain an even temperature.

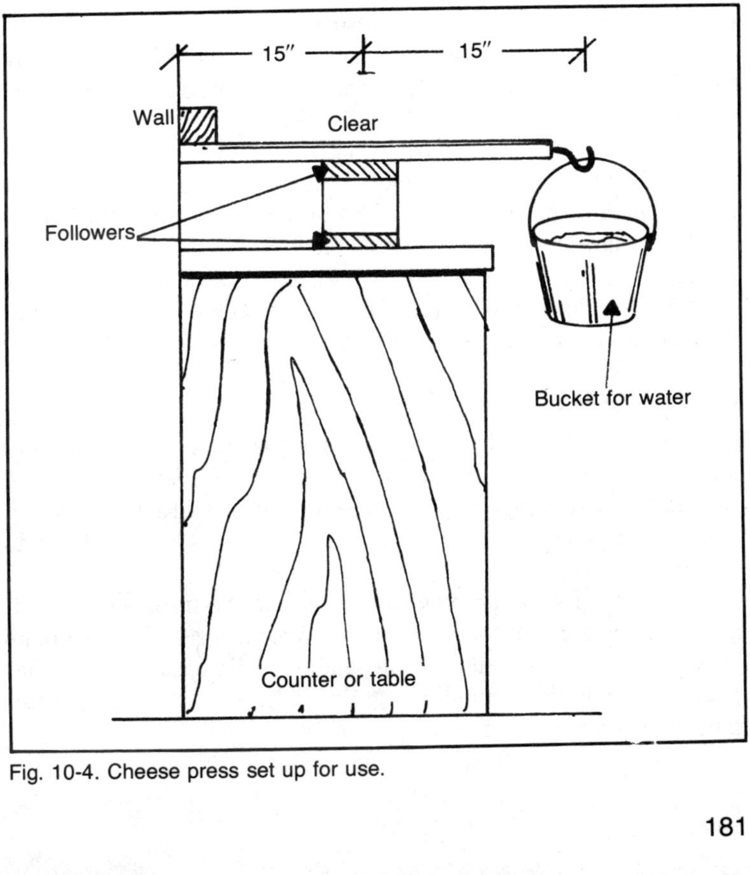

Fig. 10-4. Cheese press set up for use.

Begin by pasteurizing the milk. Prepare a starter by adding 1 tablespoon of fresh buttermilk to 1 pint of pasteurized milk. Hold this mixture at 72 degrees Fahrenheit for about 16 to 24 hours until it curdles. Scald a teaspoon, then cool it, and add a teaspoonful of the curdled milk to another pint of pasteurized milk. This second pint should curdle in about 12 hours or more. Add ¼ cup of this second pint of starter to a gallon of pasteurized milk at 72 degrees Fahrenheit. Mix the starter into the milk, then don't stir the milk again. Maintain the temperature at 72 degrees Fahrenheit until the milk curdles. The curdling time should be from 16 to 24 hours. If the time is shorter than that, add less starter the next time you make cheese. If it is longer than 24 hours, add more starter the next time. The curd is ready when it breaks cleanly.

Cut the curd by using the same technique described earlier for making Colby cheese. Heat the curd slowly, bringing the temperature up from 72 degrees to 100 degrees Fahrenheit at a rate of about 1 degree per minute. Stir the curd every few minutes to keep the temperature uniform through the curd. After the curd reaches 100 degrees Fahrenheit, you can heat it faster with frequent stirring. When it reaches 115 degrees Fahrenheit, hold it at that temperature for 20 to 30 minutes. If the curd doesn't become firm in that time, increase the temperature by 5 or 10 degrees.

When the curd is firm, bail out the whey. Put the curd in a cheesecloth bag and drain. Immerse the bag of curd in a container of cold water. Rinse it in ice water for a few minutes. Place the curd in a colander and leave it to drain until the draining stops. Transfer the curd to a mixing bowl and salt it, using a teaspoon of salt per pound of curd. You can leave it unsalted. Add about 5 tablespoons of sweet or sour cream to the cheese, mix, and refrigerate it.

GOAT ICE CREAM

As in many other foods, you can use goat milk as a replacement for whole cow milk. The chances are that your goat milk will add richness to any recipe, because most recipes will have been developed for commercial milk. The fat content has been reduced to the legal limit.

To make ice cream, you can either use the freezing compartment in your refrigerator or get an ice cream freezer. The ice cream freezer, with its internal rotating paddles, will produce a smoother product. If you don't mind a slightly icy texture, the refrigerator method works just fine.

Refrigerator Sherbet

Here is a recipe that requires few ingredients and yet is popular with most people.

- 1 quart milk
- 1 cup sugar
- ½ cup corn syrup
- 3 lemons
- 1 pinch salt

Mix the milk, sugar, corn syrup, and salt. Pour the mixture into a loaf pan or into empty ice cube trays. Place the pan in the freezing compartment of the refrigerator. Squeeze the lemons and, after the mixture is partially frozen, stir the lemon juice into the mixture. To reduce the size of the ice crystals, remove the mixture from the freezing compartment at least once more and stir it with a fork. For even smoother sherbet, use a cold bowl and a mixer to break up the ice crystals.

Ice Cream Freezer Use

To make ice cream in a home freezer, you must have a supply of finely crushed ice. If you live near an ice plant, you can buy ice in a block at a lower price than you will pay for crushed ice. An ice chipper is useful, or you can put the ice block in a canvas bag and pound it with a heavy hammer. The ice must be crushed fine so it will melt quickly. Freezing is accomplished by removing heat from the ice cream mixture. The change of state when ice turns to a liquid absorbs far more heat than the minor heat change accomplished by holding a piece of ice against the freezer can.

One way to handle the ice is to make a wooden box to hold it. You can crush the ice in the box, then add enough salt to make a mixture that will aid the melting process. For a gallon-size freezer you will need about 15 pounds of ice and either 3 cups of fine salt or 5 cups of rock salt. Chill the ice cream mix before starting the freezer.

Basic Ice Cream Recipe

This recipe can serve as a base for many variations.

- ½ gallon milk
- 1 pint cream
- 3 tablespoons unflavored gelatin
- 6 tablespoons cold water
- 1-½ cups sugar
- ½ cup corn syrup
- ¾ teaspoon salt
- 4 teaspoons vanilla flavoring

Soften the gelatin in 6 tablespoons of cold water. Scald 2 cups of milk and add the softened gelatin to it. Stir until the gelatin is dissolved. Add sugar, syrup, and salt. Remove from heat and add the rest of the milk, cream, and vanilla. Chill and then freeze.

Here are some variations. For chocolate ice cream, add ½ cup of cocoa to the sugar. For mocha ice cream, dissolve about 3 teaspoons of instant coffee in a small amount of warm milk in a custard cup. When the mixture is smooth, add it to the milk. For any fruit ice cream, add 2 cups of crushed fruit such as peaches or strawberries to the sugar and milk. You can omit the vanilla flavoring for fruit ice creams. You may want to add more sugar to the recipe for some fruits.

MEAT

Goat meat is held in high esteem in many parts of the world, but how often have you seen any at your supermarket? Hunters invest heavily in clothing, weapons, and vehicles to hunt deer to get venison, which has a taste similar to goat meat. The venison is often handled in the field in a way that would turn the stomach of a carrion eater. One meat processor told me he always dreads deer season because of mishandled carcasses he has to process, starting with intestinal material spread through the meat by bullets and complicated by trash picked up while the carcass was dragged through the woods. Even if you have never butchered anything, you can do a better job on your goats than the typical hunter does with deer.

Goat meat can vary in flavor and texture, depending on how old you let the kid get before you kill it. Many people like milk-fed animals, so they butcher kids well before weaning time. This makes for more tender, milder tasting meat. The meat yield per animal is limited. If you have only a single doe, you probably should let your kids grow to at least six months old before butchering. From a six-month-old kid you can expect to get about 30 pounds of meat. If you remove the meat from the bones, the net weight will be much less than 30 pounds.

Before killing a goat, prepare for proper handling of the carcass. You will need a place to hang the animal while skinning him, either a well supported beam or a tree limb (Fig. 10-5). A pulley attached to the tree limb allows you to pull the carcass up without help. An even better alternative is to get or borrow a fence stretcher of the block and tackle variety. You can easily pull the carcass up with it and the catch on the stretcher will hold the rope at any point you choose.

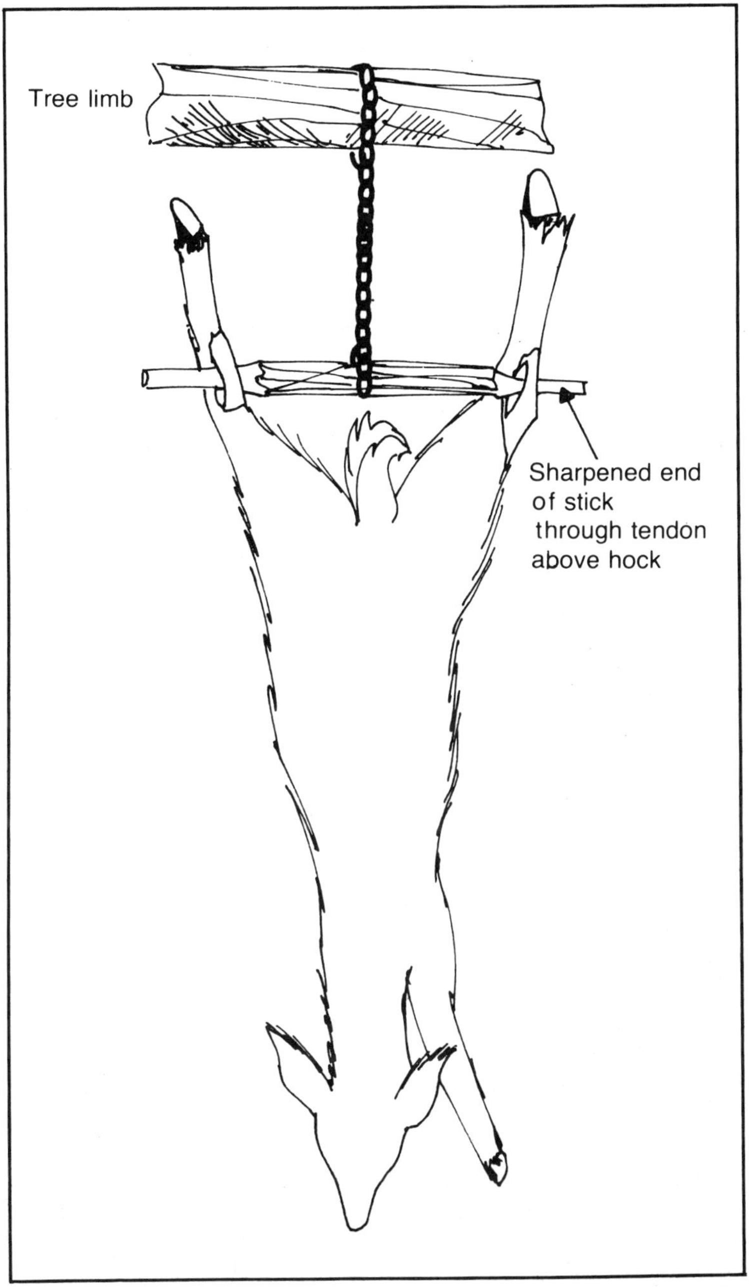

Fig. 10-5. How to use a hanging stick and wire or chain to hang a goat carcass for skinning and opening the carcass.

To support the carcass and hold the rear legs apart, prepare a stick to insert through the rear legs (Fig. 10-6). This substitutes for the single tree with hooks used by professional butchers.The stick should be strong enough for any animal you want to kill; one with a diameter of about 2″ should be adequate.

Butchering will be easier if you starve the animal for at least 12 to 24 hours before killing. This empties the rumen and other internal organs of the digestive tract, making them easier to handle. Don't do anything to upset the kid that you will kill. If you put him in solitary confinement, he may get so agitated that the meat will be affected. Try to make arrangements for him to have the company of another goat during the starvation period.

Sharpen several knives. A meat saw is almost a necessity, but you can use any hacksaw just as easily. If you use a hacksaw, make sure it has a blade with no paint on it. Clean the blade thoroughly before converting it to a meat saw.

You will need two buckets of clean water. One is for hand washing, and the other is used to receive the liver and heart. A 12-gauge shotgun is needed to quickly kill the animal.

Don't plan to kill several goats at once, unless you have been through the butchering process several times. Home freezers are adequate to handle a small amount of unfrozen meat. If you add too much at once, you will raise the temperature of the frozen foods and lower their flavor and keeping quality. You should ideally kill no more than a single goat at a time.

Plan your work carefully so that the killing, skinning and other work is done at some distance from any odors such as those of manure, gasoline, or paint. Meat can absorb these odors, impairing the flavor. Wear clean clothing and handle the skinned carcass by either carrying it or lowering it onto a clean sheet to protect it from the ground.

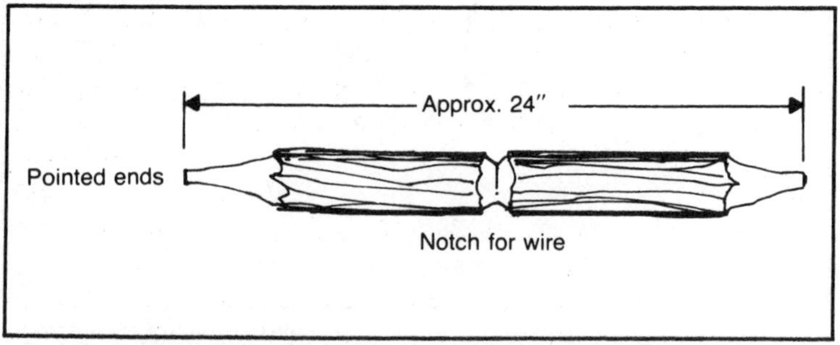

Fig. 10-6. Approximate dimensions for a hanging stick.

KILLING METHODS

You should try to kill the animal as humanely as possible, and goats require special methods. Even experienced butchers sometimes have difficulty making a clean kill with goats because they lack experience with them. One man I know tried to use his favorite method of killing calves, a stunning blow to the forehead with a ball peen hammer, on a goat. It didn't work.

For a quick, humane, stunning blow, use a 12-gauge shotgun and aim at the back of the head. A shotgun has several advantages over the popular 22-caliber rifle. It is safer to use in an inhabited neighborhood. The total mass of the load is greater, having more stunning power.

As soon as the goat drops, slit its throat with a sharp knife. Make sure you cut the jugular. Some people prefer to have the goat hanging within seconds, but this seems unnecessary.

SKINNING

With the carcass on the ground, make cuts around the legs just below the hocks. Use enough pressure to cut through the skin. After circling the leg, insert the point of the knife blade under the skin on the inside of a leg. Run the blade toward the crotch. The sharp edge of the blade should be turned outward toward the skin, away from the muscle, to split the skin along the inside of the leg. Grasp the corner of the skin made by the circular cut and the cut along the leg. Pull the skin loose. Slip the knife under the skin to slice it loose from the muscles under it. This is the technique you will use throughout the skinning process.

As soon as you have bared the hock area on both rear legs, make a slit through each leg above the hock. Insert the knife blade between the large tendon at the back of the leg and the bone. Insert the points at the ends of the hanging stick through these slits (Fig.10-7). Hang the animal for easy skinning. Hoist it until the head clears the ground.

Split the hide from the area of the rear legs down the front of the belly to the throat. Take care that the knife doesn't go too deeply. Remember to pull the skin away from the carcass with one hand as you slice it loose with the knife (Fig. 10-8). Cut around the forelegs as you did the rear legs. Make slits along the inside of the forelegs to remove the skin there. Use a meat saw to remove the forelegs at the knees.

There are two general methods of removing the internal organs. Some people prefer to open the carcass and let the organs fall

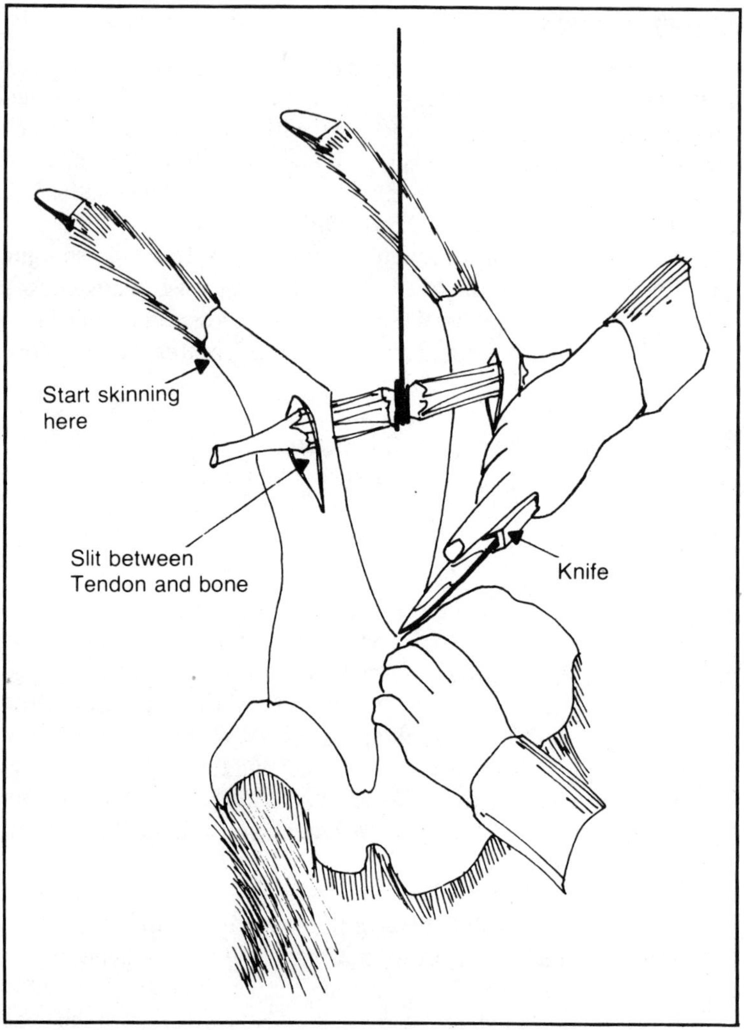

Fig. 10-7. How to attach the hanging stick to the rear legs.

over the front of the brisket. Others like to saw through the brisket so that it will be open and let the internal organs fall down through the open carcass. Regardless of which method you use, cut off the head after skinning the animal. Use your meat saw to cut it at the base of the skull.

Go back to the tail area and cut around the anus. Cut and pull it and the large intestine loose from the backbone. For safety while handling it, tie it with a cord. Open the body cavity along the front.

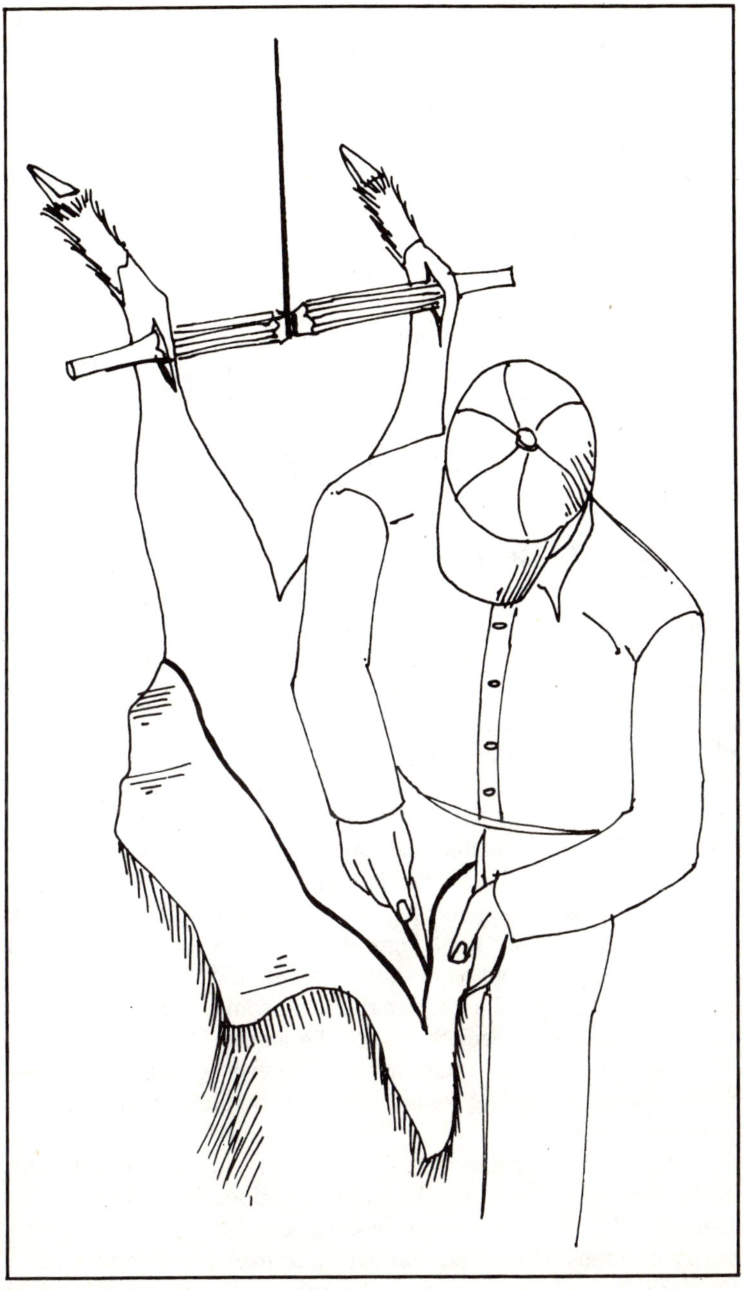

Fig. 10-8. Pull the skin away with one hand while you slice it free from the carcass with a knife.

Use care to cut around the penis to loosen it and the attached organs from the carcass. Don't cut through the ureter; an accident there could spill urine into the body cavity.

Continue to cut open the carcass down to the brisket. Pull the internal organs loose so that they fall out and hang there. Remove the liver. Locate the gall bladder, which is a dark gland attached to the liver. Cut around it so that you remove some liver on all sides of it, then discard the gall bladder. Use care in handling this organ; if you puncture it the bile will flavor the liver. Drop the liver into a bucket of cold water.

Locate the esophagus where it goes through the diaphragm and cut it there, allowing the intestines to fall to the ground. Open the brisket with your meat saw (Fig. 10-9); then remove the heart and lungs. Put the heart into cold water with the liver. Clean any stray tissue from the interior of the carcass. Splash a bucket of cold water into the carcass to clean it of any bits of blood or other material.

CUTTING THE CARCASS

The traditional way to handle any carcass is to split it down the middle with a meat saw, cutting through the backbone. If you want to follow traditional methods of cutting meat, get a bulletin on lamb processing from your local county agent's office. The method to be described here is unorthodox, but it saves freezer space by eliminating most of the bone. With this method you will have only three or four kinds of meat: whole legs, boneless stew meat, ribs, and possibly goatburger. If you want steaks or chops, get the bulletins that are available and follow them. Remember that goat meat has less marbling than other meats. This makes steaks and chops rather tough, especially because home aging is impractical. Beef is normally hung at 40 degrees Fahrenheit for at least several days before cutting, but home-killed goat should be cut immediately and frozen. Aging will tenderize the meat somewhat, but the flavor is often impaired, especially where the animal is handled at home under less than sterile conditions. To properly age meat, the conditions of aging, including temperature and humidity, must be carefully controlled.

Place the carcass on a clean table. You don't have to cut it into halves for the cutting process to be described here. Remove the legs first. These make good roasts. If you want to use them for that purpose, wrap and freeze them. Use a freezer paper to wrap meat for freezing. Pull it tightly around the meat so that the air around the meat will be expelled. Fold the ends of the paper and seal it with

Fig. 10-9. Cutting through the brisket after removing the skin and intestines.

freezer tape. Place the meat loosely in the freezer, so there is plenty of air space around each piece.

After taking care of the legs, use the meat saw to cut off the ribs from the backbone region (Fig. 10-10). Because the ribs occupy considerable space for the quality of meat they hold, use them immediately or within a day or so rather than freeze them.

Use your knife to remove chunks of meat from either side of the backbone. Cut these pieces into 1″ cubes. Freeze them in packages of convenient size for your family. If you are limited for freezer

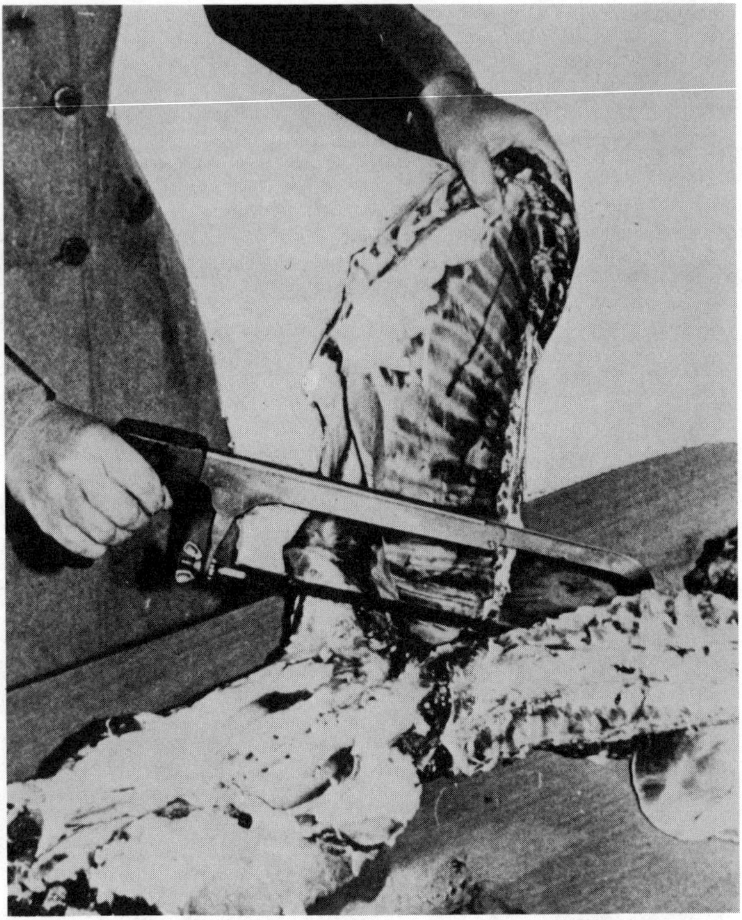

Fig. 10-10. Use a clean meat saw or a hacksaw to cut the ribs from the backbone.

space, you can make stew of the legs. Continue to work over the carcass. Remove any lean meat that you find for stew meat. You can grind some of the meat into goatburger as an alternative. When I butcher a goat, I usually save one or two legs for roasts. The rest of the carcass goes into stew meat. This method eliminates nearly all the bone from the freezer and gives me goat meat prepared the way I like it best. You can adapt the cutting process to your own needs.

RECIPES

You can obviously substitute goat meat for other meats in a multitude of standard recipes. Here are a few that seem to work especially well with goat meat. If you like soups, make stock from the bones before you discard them.

Soup Bones and Soup Stock

After you have cut the excess meat from the bones, use the bones to make soup stock. Put the bones into a large pan or several large pans,cover with water, and bring to a boil. Allow the bones to simmer for a couple of hours or more to make stock. Sometimes the aromas from this process may seem unappetizing, but don't worry about that. The stock will make wonderful soups.

When the stock has cooled, remove the bones and strip any bits of meat from them. Pour the stock and meat into freezer jars. Make sure the jars sit upright in your freezer, so that you can cut the accumulated fat off the top of the stock when you remove a jar from the freezer. If you like all that fat in your soups leave it. You'll find that this stock improves almost any soup.

Barbecued Ribs

After removing the ribs from the carcass (Fig. 10-10), cut the ribs apart. Leave about four or five ribs in each piece. Place the ribs in a roasting pan with a little water and bake uncovered for one or two hours at 200 to 250 degrees Fahrenheit. Remove the ribs from the oven and drain off the liquid. Coat both sides of each piece of ribs with a commercial barbecue sauce. Cover and bake for two to three hours at 300 degrees Fahrenheit. You may have to add a little water during the baking to keep them from sticking. To test the meat to see if it is done, wiggle a rib. The bones should almost fall out when the meat is thoroughly cooked.

Roast Leg of Goat

- 1 leg of goat
- 1 tablespoon salt
- ¼ teaspoon pepper
- ½ cup lemon juice
- 1 cup of vinegar
- ½ cup of melted butter

Rub the meat with salt and pepper. Mix the lemon juice and vinegar, then pour the mixture into the bottom of the roasting pan. Place the leg of goat in the pan and marinate for more than an hour. Turn the meat frequently. At the time of roasting, combine the butter with the liquid in the roasting pan. Baste the roast frequently with the mixture. Set the oven at 450 degrees Fahrenheit for the first 30 minutes then reduce the temperature to 300 degrees Fahrenheit for the rest of the time. A typical goat leg will roast in about 90 minutes. If you have scales, weigh the leg and allow about

20 to 25 minutes per pound. You can use wine instead of vinegar for a more aromatic roast. Either wine or vinegar will act as a tenderizing agent. The alcohol cooks out with wine, leaving a somewhat fruity gravy that many people enjoy.

Spanish Stew

1	pound cubed goat meat	1	cup diced celery
3	slices diced bacon	½	teaspoon poultry seasoning
1	large chopped onion	1-½	teaspoon vinegar
3	sliced carrots	¼	cup chopped green pepper
1	cup canned tomatoes		salt to taste

Try to get good hickory-smoked bacon when collecting materials for this stew. For poultry seasoning choose a brand that lists thyme first in the list of ingredients. You can substitute ¼ teaspoon of thyme and ¼ teaspoon of sage for the poultry seasoning.

Fry the bacon until it is browned. Brown the goat meat in the bacon fat and salt lightly. Add onion and cook for 10 minutes. Add other ingredients, cover, and simmer for an hour. Serve over rice.

Scotch Stew

1	pound cubed goat meat	1	cup diced cauliflower
1	quart water	¼	cup sliced green onions
1	cup diced carrots		salt and pepper to taste
1	cup diced turnips	2	stalks diced celery
1	cup green peas	¼	cup chopped parsley

Brown the meat in a saucepan. Then cover with water and bring to a boil. Simmer for an hour. Add the turnips, carrots, green onions, salt, and pepper. Cook for another hour. Add the green peas and cauliflower. Cook until they are tender. Add the parsley and serve.

Many variations are possible with this stew. You can add almost any vegetable to vary the taste.

Italian Stew

2	pounds cubed goat meat	1	can mushrooms
1-2	cloves garlic	1	tablespoon rosemary
2	cups red wine	3	chopped green peppers
1	large chopped onion		salt and pepper

Soak the goat in wine for several hours. Brown meat, onion, and garlic in oil. Cover and cook at low heat for one hour. Add other ingredients and cook for another hour or more.

Goatburger and Its Uses

If you have a meat grinder, you can make goatburger that will serve in any recipe that calls for hamburger. For hamburgers with a unique taste, mix some goatburger with your beef hamburger. Another possibility is to make your goatburger into sausage. To make goat sausage, just add some poultry seasoning to the meat and work it in with your fingers until it is well mixed. You can do this before freezing the meat, but ideally the spices should be added just before cooking.

Goatburger makes an excellent chili. Just brown the meat, then add it to your favorite chili recipe.

How about goatburger and spaghetti? You can use goatburger in many casserole dishes. Here is one that you can make with either a slow cooker or in your oven.

Goatburger and Noodle Casserole

- 1 pound goatburger
- 1 8 to 10-ounce package noodles
- 1 large chopped onion
- 2 stalks chopped celery
- 1 6-ounce can tomato paste
- 1-¼ cups of water
- 1 10-¾-ounce can tomato soup
- 1 teaspoon leaf oregano
- salt and pepper to taste

Cook the noodles, following the directions on the package. If the noodles tend to stick together, run cold water through them to separate them. Brown the goatburger. Place the noodles in an oiled slow cooker if one is available. If not, place in an oiled casserole for oven cooking. Add the remaining ingredients and stir to mix. Cover and cook on low heat in a slow cooker for five to eight hours. If you are cooking this casserole in an oven, set the oven at 300 degrees Fahrenheit and cook for two or three hours. Set the oven at 200 degrees Fahrenheit and cook for five hours.

GOATS AND HOME ECONOMICS

The uses of goat milk and meat mentioned in this chapter are just a few ways that goats can reduce your food bills. People who raise pigs like to boast about how much meat each one yields. Several goat kids are needed to equal a single pig in meat production, but you can't raise a pig on little more than milk and browse. As a pig reaches the age of butchering, it usually consumes costly feed at an alarming rate. Depression farmers called the pig a mortgage lifter. I nominate the goat as the budget saver in today's world.

A writer once suggested that he considered the goat a harbinger of poverty because goats are kept in underdeveloped countries. This image of the goat may make some insecure people hesitate to keep goats, and it may explain why goat keepers are often strongly individualistic. Linking goats with poverty is akin to saying that bicycles cause poverty, and that if poor people would drive Mercedes Benz cars, their financial problems would be solved. In the real world it is the other way around. Besides, costly trinkets aren't necessary for the really secure. If owning goats is nose-thumbing to the establishment, so be it. A Mercedes is okay if you can afford it. When you go out to admire the car, it won't frolic across the meadow to meet you.

Appendix

Goat Associations, Magazines, and Supply Sources

DAIRY GOAT ASSOCIATIONS

American Dairy Goat Association (ADGA)
P.O. Box 865
Spindale, NC 28160

American Goat Society (AGS)
Rt. 2, Box 112
DeLeon, TX 76444

GOAT PERIODICALS

Dairy Goat Guide
312 Portland Road
Highway 19 East
Waterloo, WI 53594

Dairy Goat Journal
Box 1808
Scottsdale, AZ 85252

United Caprine News
P.O. Drawer A
Rotan, TX 79546

Caprine Supply
6657 Woodland
Shawnee, KS 66218

An excellent source of general supplies. The entire catalog is slanted to the goat keeper.

Countryside General Store
103 North Monroe Street
Waterloo, WI 53594

Havasu Hills
704 "I" Street
Modesto, CA 95354

Leather collars

Hoegger Supply Co.
P.O. Box 331
Fayetteville, GA 30214

New England Cheesemaking Supply Co.
Box 8503
Ashfield, MA 01330

Pet-Set Nylon Supply
311 E. Wallen Dr.
Brooksville, FL 33512

Nylon collars, etc.

Index

Edited by Bob Ostrander